How to Recognize a Hazardous Waste

(even if it's wearing dark glasses)

A Concise Training and
Reference Guide
for RCRA Compliance

Gary Crouth

Illustrations by William M. Vrscak

How to Recognize a Hazardous Waste (even if it's wearing dark glasses):
A Concise Training and Reference Guide for RCRA Compliance

Copyright © 1987, 2012 by OLAP World Press
First edition published 1987. Twelfth edition 2012.

ISBN-13: 978-0-9817753-2-6
ISBN-10: 0-9817753-2-2

OLAP World Press
114 Elysian Street
Pittsburgh, PA 15206
info@OLAPWorldPress.com
http://OLAPWorldPress.com/HazardousWaste12thEdition

Formerly published by Digby Books Ltd.

Illustrations by William M. Vrscak

21 20 19 18 17 16 15 14 13 12   1 2 3 4 5

CONTENTS

FIGURES AND TABLES

Appendix 1: RCRA Hazardous Wastes Lists

PREFACE

I MAGINE YOURSELF ASSIGNED TO a mission in a far-off land. Upon arrival you are immersed in a culture whose language uses familiar words with foreign meanings, where commonsense must be abandoned. You find yourself entangled in a web of definitions, policies, rules, and guidelines that are ambiguous, unfamiliar, and sometimes contradictory. It is possible to become stuck there, in frustration and confusion, never quite succeeding in the mission, yet unable to give it up. Your mission is to track down and identify those materials regulated as hazardous wastes by the law of the land.

Now in its twelfth edition, this book is a well-tested travel guide to the strange world of the Resource Conservation and Recovery Act (RCRA) hazardous waste characterization. Just as a travel guide is no substitute for traveling, this book is by no means a substitute for the regulations. The goal of this book is to orient you, focus your efforts, and guide you through the regulatory framework of definitions and criteria, so you can properly identify RCRA hazardous wastes.

Characterization of hazardous wastes is the first crucial step in any waste management process. RCRA hazardous wastes require very special management to protect yourself from the long, sharp teeth of the RCRA enforcement provisions and to help prevent environmental problems from waste mismanagement.

Keep in mind that the word "hazardous" is strictly the RCRA regulatory definition that is based on narrowly defined criteria, sweeping generic listings, and rather arbitrary rules. As a result, some wastes are classified as hazardous that do not present much of an environmental threat relative to other wastes. Likewise, there are wastes that escape the hazardous label under this system that present serious hazards and should be managed carefully. *Be aware of this paradox.*

Throughout the book, the applicable sections of the *Code of Federal Regulations* (CFR) are noted in the left margin. This book is based on federal EPA regulations, guidelines, and policies, and it is applicable in most states. Some states have established criteria for identifying hazardous wastes that surpass the minimum set by federal regulation, and there are some wastes the U.S. EPA has added to the hazardous waste list that have yet to be adopted by some states. Also, some regulations are still being adopted by the states. Therefore, you are best served by having up-to-date bookmarks of both federal and state regulations in your web browser. The URLs in APPENDIX 2: *For Further Information* and APPENDIX 3: *Significant Regulatory Memos* are a good place to begin.

ACKNOWLEDGEMENTS

No one does anything alone, but only in cooperation with countless others. I feel deep gratitude for all who in any way helped bring this book into being, but here are some I wish to thank more explicitly.

John Paredes of OLAP World Press, my publisher, through his vision and understanding of and appreciation for this book has enabled it to have a renewed life.

Jill Cooper, whose meticulous editing and attention to detail corrected and clarified much information, and whose formatting expertise has given this book its polished and professional look.

Deasy Suryani for designing such an eye-appealing cover.

Professor Jeanette Garr of Youngstown State University, who urged expansion of the book and contributed ideas and information included in CHAPTER 8.

Bill Vrscak drew the illustrations that offer harmonious and whimsical accompaniment to my words.

Jenny Wolsk Bain gave me encouragement, smiles, and comfortable office space that helped me finish writing the review questions and answers and CHAPTER 8.

My parents, Joe and Joanne Crouth, and daughters, Jasmin and Brennan, whose love and support flow through everything I do, including this book.

My former colleagues in the environmental business provided me the opportunities for learning what I've written in this book.

Deep bows to all my teachers across the years who continue to light the way.

INTRODUCTION

THE FIRST FEDERAL SOLID waste law, the Solid Waste Disposal Act, was passed in 1965. In 1976, Congress amended this law by replacing its language entirely with the Resource Conservation and Recovery Act, commonly known as RCRA (pronounced "wreck-rah"). RCRA established the framework for managing both solid and hazardous waste. This framework consists of ten subtitles: A through J. Most of the subtitles establish the federal bureaucracy for administering the RCRA regulatory program. Subtitle C gives the U.S. EPA authority and directives to develop the hazardous waste regulatory system for the U.S. Most of the hazardous waste regulations were born from the following sections of the statute:

- 3001 Identification and listing of hazardous waste
- 3002 Standards applicable to generators of hazardous waste
- 3003 Standards applicable to transporters of hazardous waste
- 3004 Standards applicable to owners and operators of hazardous waste treatment, storage, and disposal facilities
- 3005 Permits for treatment, storage, or disposal of hazardous waste

EPA published the first batch of hazardous waste regulations on May 19, 1980, which became effective on November 19, 1980. Since then, these regulations have been amended many times, and they are still changing.

This book refers to the regulations EPA has promulgated from Section 3001 of Subtitle C of RCRA, and it deals primarily with the question, What is a "hazardous waste" as defined and regulated by RCRA?

CHAPTER 1: *Solid Wastes* discusses the concept of "RCRA solid waste" and the explicit exemptions that exclude certain materials from being defined as solid wastes and certain solid wastes from being defined as hazardous wastes. Solid wastes regulated by laws other than RCRA are also briefly discussed. Each chapter has a set of review Q&A to help you assess your understanding of the material.

CHAPTER 2: *Hazardous Waste Definition* is the heart of this book. It offers a systematic thought process for determining if a solid waste meets the RCRA definition of hazardous. First, you are introduced to the three general categories EPA uses to determine if a waste is hazardous, as well as the six EPA Hazard Codes. Then you are guided through the three hazardous waste lists. Then the four hazardous waste characteristics are explained. Next, we deal with how to classify hazardous wastes. Lastly, testing wastes and documentation requirements are discussed.

CHAPTER 3: *Used Oil* discusses the very special status of used oils under RCRA. A series of six questions is provided: use questions 1–3 to classify a used oil under RCRA, use questions 1–5 if you burn used oil or send it directly to a burner, and use questions 1–6 if you claim that used oil is "specification used oil fuel." Once you have answered the questions that apply to your situation, read through the discussion of the regulations that apply to your role in regard to used oil. The final discussion is of miscellaneous regulations about prohibitions, exemptions, and PCBs.

CHAPTER 4: *Other Waste Characterization Principles* reviews some principles of waste identification: Empty Container Rule, Mixture Rule, Derived-From Rule, Contained-In Policy, Contained-In Rule, and the characterization of spill residues.

CHAPTER 5: *Universal Wastes* deals with identifying universal wastes: those common items that become hazardous wastes when discarded. The distinction between small and large quantity handlers is discussed as well as the requirements for each.

CHAPTER 6: *Hazardous Wastes Used, Reused, Reclaimed, or Recycled* provides a six-step process to determine which hazardous wastes are not considered wastes and so are exempted from RCRA because they are used, reused, reclaimed, or recycled—a very narrow exemption. Waste vs. raw material is discussed, and the criteria for non-waste determinations is reviewed.

CHAPTER 7: *Delisting* briefly explains the delisting mechanism—how it works, when it works, and how to determine if your wastes may be delistable.

CHAPTER 8: *Additional Tips and Advice for Accurate Waste Recognition* provides specific guidance on using CAS numbers and MSDSs to help you accurately identify hazardous wastes. The point in time when a material becomes a hazardous waste is explained. Hazardous terms are reviewed.

APPENDIX 1: *RCRA Hazardous Wastes Lists* are current as of the book's printing date. Included are *Non-specific Source Wastes* (§261.31), *Specific Source Wastes* (§261.32), and *Discarded Commercial Chemicals Products:* Acutely Hazardous [§261.33(e)] and Hazardous [§261.33(f)].

APPENDIX 2: *For Further Information* lists sources of information that may be useful or necessary to properly characterize and classify your hazardous wastes.

APPENDIX 3: *Significant Regulatory Memos* is a list of regulatory memos written by U.S. EPA officials that offer clarification and insight into EPA's interpretation of RCRA regulations.

ABBREVIATIONS

API	American Petroleum Institute
ASTM	ASTM International (pre-2001 the American Society for Testing and Materials)
atm	atmosphere (measure of pressure)
(C)	EPA Hazard Code: corrosive
CAS	Chemical Abstracts Service
CERCLA	Comprehensive Environmental Response, Compensation, and Liability Act (aka Superfund)
CFC	chlorofluorocarbon
CFR	*Code of Federal Regulations*
CRT	cathode ray tube
DAF	dilution/attenuation factor
DAF	dissolved air flotation
DDD	dichloro-diphenyl-dichloroethane
DDT	dichloro-diphenyl-trichloroethane
DFP	diisopropylfluorophosphate
DIY	do-it-yourself
DOT	Department of Transportation
DRAS	Delisting Risk Assessment Software
(E)	EPA Hazard Code: toxicity characteristic
EPA	Environmental Protection Agency
EPACMTP	EPA Composite Model for Leachate Migration with Transformation Products
EPCRA	Emergency Planning and Community Right-to-Know Act
°F	degree Fahrenheit
FR	*Federal Register*
(H)	EPA Hazard Code: acute hazardous
(I)	EPA Hazard Code: ignitable
IAF	induced air flotation
kg	kilogram

l	liter
LDR	land disposal restriction
LQH	large quantity handler
MCL	maximum contaminant level
mg	milligram
mg/kg	milligrams per kilogram
mg/l	milligrams per liter
ml	milliliter
mm	millimeter
MNNG	1-methyl-2-nitro-1-nitrosoguanidine
MSDS	material safety data sheet
MSMA	monosodium methyl arsenate
OSHA	Occupational Safety and Health Administration
PCB	polychlorinated biphenyl
PCNB	pentachloronitrobenzene
pH	power of Hydrogen (measure of acidity)
POTW	publicly owned treatment works
ppm	parts per million
psi	pounds per square inch
(R)	EPA Hazard Code: reactive
RCRA	Resource Conservation and Recovery Act
SARA	Superfund Amendments and Reauthorization Act
SIC	Standard Industrial Classification
SQH	small quantity handler
(T)	EPA Hazard Code: toxic
TC	toxicity characteristic
TCE	trichloroethylene
TCLP	Toxicity Characteristic Leaching Procedure
TNT	trinitrotoluene (2,4,6-trinitrotoluene)
TSCA	Toxic Substances Control Act
UDMH	unsymmetrical dimethylhydrazine (1,1-dimethylhydrazine)
URL	Uniform Resource Locator

CHAPTER 1:
Solid Wastes

A s we begin the process of recognizing hazardous waste, please dismiss any notion that "solid waste" refers to physical state: a RCRA solid waste can be solid, liquid, or containerized gas.

Solid Wastes Identification

§261.2

A material must first be a "solid waste" before it can be a "hazardous waste." A material is "discarded" and therefore a solid waste if it is one of the following four things:

1. Abandoned by being disposed of; burned or incinerated; or accumulated, stored, or treated instead of or prior to being disposed of, burned, or incinerated.

   By *abandoned* EPA has explained: "We do not intend any complicated concept, but simply mean thrown away." [1]

---

1. 50 FR 627, *Federal Register*, January 4, 1985, p. 627.

2. Recycled in certain ways as spent material, sludge, by-product, commercial chemical product, hazardous scrap metal, or excluded scrap metal.

   Determining when a recycled material is a solid waste rather than a raw material is a complex ordeal discussed in CHAPTER 6: *Hazardous Wastes Used, Reused, Reclaimed, or Recycled*, page 71.

3. Listed as "inherently waste-like."

   "Inherently waste-like" is simply a material on the "inherently waste-like" list. See CHAPTER 6: *Hazardous Wastes Used, Reused, Reclaimed, or Recycled*, page 73, for the full discussion.

4. Military munition identified as solid waste.

   Special regulations identifying military munitions as hazardous wastes are at 40 CFR 266.202.

Specifically Exempted Wastes

In §261.4(a) there is a list of materials that are not solid wastes, and in §261.4(b) there is a list of solid wastes that are not hazardous wastes. There are many exclusions, and only the most common ones are listed here. The exclusions are very specific, so if you see an exclusion that you believe exempts your waste, read the regulatory language to fully understand the exemption.

§261.4(a)

MATERIALS THAT ARE NOT SOLID WASTES

- Untreated sanitary sewage and mixtures of wastes with sanitary sewage sent to a publicly owned treatment works (POTW).

- Point source discharges regulated by the Clean Water Act.

  Exemption is at the point of discharge. For example, exemption does not apply to material while being stored in a surface impoundment prior to discharge.

- Irrigation return flows.

- Scrap metal that has been processed such as baled, shredded, sheared, chopped, crushed, flattened, cut, melted, or sorted.

- Metal fines or drosses that have been agglomerated.

- Metal turnings, cuttings, punchings, and borings from metal mills, foundaries, refineries, and metal working/fabricating plants.

- Used cathode ray tubes (CRTs).

- Hazardous secondary material generated and reclaimed within the U.S. or its territories and managed in land-based units.

- Hazardous secondary material that is generated and then transferred to another person for the purpose of reclamation.

There are other *very specific* exemptions for certain nuclear wastes, in-situ mining wastes, pulping liquors, spent sulfuric acids, reclaimed spent wood preserving solutions, coking and iron/steel wastes, recovered oils, shredded circuit boards to be recycled, synfuels, primary mineral processing, and petrochemical wastes. Also exempted are secondary materials meeting certain conditions that are used to make zinc fertilizers. Although not explicitly exempted, gaseous emissions, such as in a stack, are not considered solid wastes as explained by EPA at 47 FR 27530.[2]

§261.4(b)

Solid Wastes That Are Not Hazardous Wastes

- Household waste.

- Wastes from growing crops and raising animals if used as fertilizer.

- Mining overburden returned to the mine site.

- Fly ash, bottom ash, slag, and flue gas emission control waste from combustion of coal and other fossil fuels.

- Wastes from exploration, development, or production of crude oil, natural gas, or geothermal energy.

- Certain solid wastes from the extraction, beneficiation, and processing of ores and minerals (sometimes known as "Bevill exempted wastes").

   Solid wastes from "beneficiation" are defined by *the process that generated the wastes* (see TABLE 1-1, page 4). Solid wastes from "processing" are *specifically listed by industry and solid waste* (see TABLE 1-2, page 4).

- Cement kiln dust waste from kilns that process at least 50% normal cement-production raw materials.

---

2. 47 FR 27530, *Federal Register*, June 24, 1982, p. 27530.

There are also exemptions for trivalent chromium wastes from the leather industry; used refrigerants being reclaimed; discarded treated wood; leachate or gas condensate collected from landfills; and samples shipped for analysis and treatability work.

§261.4(b)(7)

Ore and Mineral Wastes Excluded

Table 1-1. Ore and Mineral Wastes Excluded Under §261.4(b)(7): "Beneficiation" Wastes

Crushing	Drying	Ion exchange
Grinding	Sintering	Solvent extraction
Washing	Pelletizing	Electrowinning
Dissolution	Briquetting	Precipitation
Crystallization	Gravity concentration	Amalgamation
Filtration	Magnetic separation	Calcining to remove water and/or carbon dioxide
Sorting	Electrostatic separation	Heap, dump, vat, tank, and in-situ leaching
Sizing	Flotation	Roasting, autoclaving, or chlorination in preparation for leaching (except where the process/leaching sequence produces a final or intermediate product that does not undergo further beneficiation or processing)

Table 1-2. Ore and Mineral Wastes Excluded Under §261.4(b)(7): "Processing" Wastes

Industry	Solid Waste
Alumina	■ Red and brown muds from bauxite refining
Chromium	■ Treated residue from roasting/leaching of chrome ore
Coal Gasification	■ Gasifier ash from coal gasification ■ Process wastewater from coal gasification

Table 1-2. (*continued*)

Industry	Solid Waste
Primary Copper	■ Slag from primary copper processing ■ Calcium sulfate wastewater treatment plant sludge from primary copper processing ■ Slag tailings from primary copper processing
Elemental Phosphorus	■ Slag from elemental phosphorus production
Ferrous Metals	■ Air pollution control dust/sludge from iron blast furnaces ■ Iron blast furnace slag ■ Basic oxygen furnace and open hearth furnace air pollution control dust/sludge from carbon steel production ■ Basic oxygen furnace and open hearth furnace slag from carbon steel production
Hydrofluoric Acid	■ Fluorogypsum from hydrofluoric acid production ■ Process wastewater from hydrofluoric acid production
Primary Lead	■ Slag from primary lead processing
Primary Magnesium	■ Process wastewater from primary magnesium processing by the anhydrous process
Phosphoric Acid	■ Phosphogypsum from phosphoric acid production ■ Process wastewater from phosphoric acid production
Titanium Tetrachloride	■ Chloride process waste solids from titanium tetrachloride production
Primary Zinc	■ Slag from primary zinc processing

Solid Wastes Regulated by Laws Other Than RCRA

Asbestos wastes are regulated by the U.S. EPA's regulations promulgated under the Clean Air Act, "National Emission Standards for Hazardous Air Pollutants" (40 CFR 61 Subpart M). These regulations include standards for removal, packaging, labeling, and disposal of asbestos wastes.

Wastes containing polychlorinated biphenyls (PCBs) above 50 ppm are also regulated by the Toxic Substances Control Act (TSCA) regulations: 40 CFR 761 Subpart B.

CHAPTER 1: REVIEW QUESTIONS

Questions

1. Liquid wastes cannot be "solid waste" under RCRA.

   ☐ Truth or ☐ Myth?

2. Which of the following could be solid wastes under RCRA? (Check all that apply.)

   ☐ A. point source discharge
   ☐ B. stack emission
   ☐ C. containerized gas
   ☐ D. wastewater in pipeline
   ☐ E. wastewater in lagoon

3. Waste that is specifically exempted from being a solid or hazardous waste is not subject to regulation as hazardous waste even if it poses serious health or environmental hazards.

   ☐ Truth or ☐ Myth?

4. Wastewater that eventually is discharged under a Clean Water Act permit is exempt from hazardous waste regulation under RCRA.

   ☐ Truth or ☐ Myth?

5. Any waste generated by beneficiation or processing of ores and minerals is automatically exempt from RCRA regulation.

   ☐ Truth or ☐ Myth?

Answers

1. ☑ Myth

    Liquid wastes cannot be "solid waste" under RCRA is a myth. The term "solid waste" has nothing to do with physical state. Solid waste can be solid, liquid, or containerized gas.

2. ☒ A: Point source discharge could not be solid wastes under RCRA at the point of discharge. Point source discharges are exempt under 40 CFR 261.4(a)(2).

    ☒ B: Stack emission could not be solid wastes under RCRA. Gaseous emissions are not solid wastes per 40 CFR 261.2 and 47 FR 27530.

    ☑ C: Containerized gas could be solid wastes under RCRA.

    ☑ D: Wastewater in pipeline could be solid wastes under RCRA.

    ☑ E: Wastewater in lagoon could be solid wastes under RCRA.

3. ☑ Truth

    Waste that is specifically exempted from being a solid or hazardous waste is not subject to regulation as hazardous waste even if it poses serious health or environmental hazards is a true statement. An exemption is not a determination about a particular waste's inherent hazards or its health or environmental risks.

4. ☑ Myth

    Wastewater that eventually is discharged under a Clean Water Act permit is exempt from hazardous waste regulation under RCRA is a myth. Wastewater is exempt only at the point of discharge. Management of wastewater prior to the point of discharge is subject to RCRA jurisdiction.

5. ☑ Myth

    Any waste generated by beneficiation or processing of ores and minerals is automatically exempt from RCRA regulation is a myth. Beneficiation wastes are exempted by process. Processing wastes are exempt if they are specifically listed. See 40 CFR 261.4(b)(7).

CHAPTER 2:
Hazardous Waste Definition

§262.11   **I**F YOUR SOLID WASTE is not specifically excluded, you are required to determine if it is a RCRA hazardous waste. There are only two ways that a solid waste may also be a RCRA hazardous waste: (1) the waste is specifically listed as hazardous, or (2) the waste fails a characteristic test.

Listed Hazardous Wastes

Under RCRA, EPA has authority to generically list wastes it believes are hazardous. If your waste meets a generic listing, the waste is automatically regulated as hazardous regardless of its properties. [1]

---

1. The waste may be "delisted" if you can successfully prove that the waste is non-hazardous according to the delisting criteria. The delisting process is described in CHAPTER 7: *Delisting*, page 91.

EPA may list a waste as hazardous if the waste "typically and frequently" falls into any of three general categories: (1) exhibits a hazardous characteristic, (2) acute hazardous, and (3) toxic.

1. **Exhibits a Hazardous Characteristic**

   The four characteristic tests are ignitability, corrosivity, reactivity, and toxicity. These tests are described in detail in *Four Hazardous Waste Characteristics*, page 16.

2. **Acute Hazardous** [2]

   The waste has been shown to be fatal to humans in low doses. EPA uses toxicity information from the literature to decide which wastes are acute hazardous.

3. **Toxic** [2]

   EPA lists a waste as "toxic" if it contains certain "hazardous constituents" and EPA judges that the waste could pose health or environmental problems according to certain criteria including chronic toxicity, migration potential, bioaccumulation, quantities of waste generated, and past environmental problems caused by mismanagement of the waste.

   Each listed waste is assigned one or more "hazard codes" that indicates the reason the waste is listed (see TABLE 2-1).

Table 2-1. EPA Hazard Codes

(I) Ignitable Waste	(E) Toxicity Characteristic Waste
(C) Corrosive Waste	(H) Acute Hazardous Waste
(R) Reactive Waste	(T) Toxic Waste

Three Hazardous Wastes Lists

There are three lists of hazardous wastes: (1) *Non-specific Source Wastes*, (2) *Specific Source Wastes*, and (3) *Discarded Commercial Chemical Products*. The lists are described in detail here and presented in full in APPENDIX 1: *RCRA Hazardous Wastes Lists*, page 107.

---

2. The entire list of "hazardous constituents" that EPA considers when listing wastes is contained in §261, Appendix VIII. For each waste listed as acute hazardous or toxic, the hazardous constituents causing that waste to be listed are given in Appendix VII.

1. NON-SPECIFIC SOURCE WASTES LIST

§261.31

The non-specific source wastes listings are broad, general groupings, each of which describes a number of different wastes generated by various industries.

Spent Solvents

F001–F005 are spent solvents and sludges from reclaiming solvents. F001 and F002 are halogenated spent solvents. F003, F004, and F005 are non-halogenated spent solvents.

Guidelines for characterizing listed spent solvents are as follows:

- Spent solvent listings regulate only those listed chemicals that have been used for their solvent properties, that is, to solubilize or mobilize other materials. Examples of this are cleaning, degreasing, and extracting.

  Wastes that contain solvents as ingredients, such as spent paint, are not regulated by these listings.

- A solvent is "spent" when it has been contaminated through use and can no longer be used without first being regenerated or reclaimed.

- A spent solvent is regulated as **F001, F002, F004,** or **F005** if before use it contained a total of 10% or more by volume of solvents in those listings. For example, a solvent blend containing 5% toluene and 8% methylene chloride is regulated as spent solvent F002/F005.

- A spent solvent is regulated as **F003** if it consists *solely* of one or more solvents listed in F003. A spent solvent waste F001, F002, F004, or F005 that contains F003-listed solvents, is also classified as F003.

- "Used for degreasing" (**F001**) means large scale degreasing operations, like vapor degreasers.

For each spent solvent listing, TABLE 2-2, page 12, gives EPA Hazardous Waste Number and Hazard Code, halogenated vs. non-halogenated information, the group of chemicals for that listing along with common chemical synonyms, and Chemical Abstracts Service (CAS) numbers.

Table 2-2. Chemicals Regulated as Listed Spent Solvents

EPA Hazardous Waste Number (Hazard Code)	Halogenated vs. Non-halogenated	Chemical	Common Chemical Synonym	CAS Number
F001 (T)	Halogenated	tetrachloroethylene	perchloroethylene	127-18-4
		trichloroethylene	trichloroethene	79-01-6
		methylene chloride	dichloromethane	75-09-2
		1,1,1-trichloroethane	methyl chloroform	71-55-6
		carbon tetrachloride	tetrachloromethane	56-23-5
		chlorinated fluorocarbons	CFCs, Freon	Various
F002 (T)	Halogenated	tetrachloroethylene	perchloroethylene	127-18-4
		trichloroethylene	trichloroethene	79-01-6
		methylene chloride	dichloromethane	75-09-2
		1,1,1-trichloroethane	methyl chloroform	71-55-6
		chlorobenzene	phenyl chloride	108-90-7
		ortho-dichlorobenzene	1,2-dichlorobenzene	95-50-1
		trichlorofluoromethane	CFC-11	75-69-4
		1,1,2-trichloroethane	ethane trichloride	79-00-5
		1,1,2-trichloro-1,2,2-trifluoroethane	CFC-113, Freon 113	76-13-1
F003 (I)	Non-halogenated	xylene	dimethyl benzene	1330-20-7
		acetone	2-propanone	67-64-1
		cyclohexanone	ketohexamethylene	108-94-1
		ethyl acetate	acetic acid, ethyl ester	141-78-6
		ethyl benzene	phenylethane	100-41-4
		ethyl ether	1,1´-oxybisethane	60-29-7
		methyl isobutyl ketone	4-methyl-2-pentanone	108-10-1
		n-butyl alcohol	1-butanol	71-36-3
		methanol	methyl alcohol	67-56-1
F004 (T)	Non-halogenated	nitrobenzene	nitrobenzol	98-95-3
		cresols/cresylic acid	methyl phenols	1319-77-3

Table 2-2. (*continued*)

EPA Hazardous Waste Number (Hazard Code)	Halogenated vs. Non-halogenated	Chemical	Common Chemical Synonym	CAS Number
F005 (I,T)	Non-halogenated	toluene	methyl benzene	108-88-3
		2-ethoxyethanol	ethylene glycol monoethyl ether	110-80-5
		methyl ethyl ketone	2-butadiene	78-93-3
		carbon disulfide	dithiocarbonic anhydride	75-15-0
		pyridine	azabenzene	110-86-1
		benzene	benzol	71-43-2
		2-nitropropane	dimethyl nitromethane	79-46-9
		isobutanol	isobutyl alcohol	78-83-1

Electroplating/Metal Finishing Wastes

F006 is wastewater treatment sludge from the following four processes:

1. Common and precious metals electroplating, *except*
   ➤ tin electroplating, *or*
   ➤ zinc electroplating (non-cyanide process, segregated from cyanide processes).

2. Anodizing, *except*
   ➤ sulfuric anodizing of aluminum.

3. Chemical etching and milling, *except*
   ➤ on aluminum.

4. Cleaning and stripping, *except*
   ➤ with tin, zinc, or aluminum plating on carbon steel.

This listing is very broad since the term "electroplating" is defined largely by EPA's pre-treatment guidelines, Part 413 – Electroplating Point Source Category. If your process falls into one of the above four categories, and that process produces a wastewater treatment sludge, that sludge is designated F006. Processes *not included* in this listing are chemical conversion coating, electroless plating, and printed circuit board manufacturing.

F019 is wastewater treatment sludge from chemical conversion coating of aluminum. The listing exempts certain wastes from zirconium phosphating in aluminum can washing and zinc phosphating of motor vehicles.

F007–F012 are hazardous waste listings for spent bath solutions and sludges from electroplating and metal heat treating operations that use cyanide in the process.

Rinsewaters from electroplating operations are not included under any of the listings. However, sludges from treatment of these rinsewaters are regulated as either F006 or F019. Also, of course, if a rinsewater exhibits one or more of the characteristics, it is a hazardous waste.

Other Non-specific Source Wastes

F020–F026 are process wastes from the production or manufacturing use of certain chlorinated organics.

F027 includes discarded unused formulations containing tri-, tetra-, or pentachlorophenols or derivatives. Some older pesticides contain chlorophenols and may be classified under this listing if discarded.

F028 is residue from incineration or thermal treatment of soil contaminated with any of the wastes F020 through F023 or F027.

F032, **F034**, and **F035** are various wastes from wood preserving.

F037 and **F038** are various wastes from petroleum refining.

F039 is leachate, and any residues from treating leachate generated from two or more wastes that are (1) listed hazardous wastes, (2) "restricted wastes" under the land disposal restriction (LDR) regulations in §268, and (3) land disposed. Leachate is liquid (including suspended solids) that has "percolated through or drained from hazardous waste" that has been either placed in a land disposal facility (for example, landfill, waste pile, or surface impoundment) or otherwise disposed of such as spilled or leaked. F039 is also called "multisource leachate."

The non-specific source wastes are assigned EPA Hazardous Waste Numbers of an "F" followed by a three-digit number.

2. Specific Source Wastes List

§261.32

Specific source wastes are specifically defined wastes that EPA has determined to be hazardous from certain industries (see Table 2-3, page 15). Each specific source listing typically describes one particular waste.

Specific source wastes are assigned EPA Hazardous Waste Numbers of "K" followed by a three-digit number.

Table 2-3. Specific Source Wastes Industries

Wood Preservation	Iron and Steel	Secondary Lead
Inorganic Pigments	Petroleum Refining	Ink Formulation
Pesticides	Explosives	Coking
Organic Chemicals	Primary Aluminum	Veterinary Pharmaceuticals
Inorganic Chemicals		

3. DISCARDED COMMERCIAL CHEMICAL PRODUCTS LIST

§261.33

A commercial chemical product or intermediate is regulated as a hazardous waste *if it is:*

➤ listed in §261.33(e)-(f) *or* its sole active ingredient is listed in §261.33(e)-(f), *and*

➤ unused, *and*

➤ this material is to be discarded for any reason (for example, it is off-spec; it has an expired shelf life; it is no longer needed; or it is contained in spill cleanup residue, debris, or contaminated media).

While some exhibit a RCRA hazardous characteristic, most of the chemicals listed in §261.33(e) are acute hazardous (H) and most of those listed in §261.33(f) are toxic (T). Empty containers of those chemicals listed in §261.33(e) are hazardous wastes unless the container's liner has been removed or the container has been decontaminated.

Note that the scope of the §261.33 (e) and (f) lists is very narrow. *This listing includes only those products and intermediates that are being discarded either in their pure form or contained in a mixture as the sole active ingredient.* Process wastes that merely contain any of the §261.33(e)-(f) chemicals—even in high concentrations—are not hazardous wastes by this listing mechanism. Also, products that contain more than one active ingredient listed in §261.33(e)-(f) are not hazardous waste by this listing.

Discarded commercial chemical products that are listed as acute hazardous in list §261.33(e) are assigned EPA Hazardous Waste Numbers of a "P" followed by a three-digit number.

Discarded commercial chemical products that are listed as hazardous in §261.33(f) are assigned EPA Hazardous Waste Numbers of a "U" followed by a three-digit number.

Four Hazardous Waste Characteristics

The next question to ask is, Does the waste exhibit one or more of the four hazardous waste characteristics?

1. **Ignitability**: Assigned EPA Hazardous Waste Number D001.

2. **Corrosivity**: Assigned EPA Hazardous Waste Number D002.

3. **Reactivity**: Assigned EPA Hazardous Waste Number D003.

4. **Toxicity**: Assigned EPA Hazardous Waste Numbers according to the maximum contaminant level (MCL) exceeded (see TABLE 2-4, page 22).

1. IGNITABILITY

§261.21

The characteristic of ignitability can apply to any physical state: solid, liquid, or containerized gas. The ignitability characteristic includes four main classes: (1) low–flash point liquids, (2) spontaneously combustible solids, (3) ignitable compressed gases, and (4) oxidizers.

Low–Flash Point Liquids

§261.21(a)(1)

The most common test of ignitability is *if the waste:*

➤ is a liquid at 68 °F and 1 atm pressure, *and*

➤ has a flash point of less than 140 °F.

Excluded from this test are aqueous liquids containing less than 24% alcohol by volume, since these liquids will often have a low flash point, but the water content keeps the liquid from supporting combustion.

Let's talk about flash point testing: Only gases burn; liquids and solids do not. However, liquids and solids can give off gases that burn. A flame is a gas that is burning, or undergoing combustion. This occurs when a gas combines with oxygen and chemically changes such that the energy in its chemical bonds releases as heat and light. A flame is sustained only as long as it gets enough combustible gases and oxygen.

A *flash point test* measures the temperature at which the vapors volatilizing from a liquid will ignite, causing the liquid itself to ignite. A solid that contains a low–flash point liquid will often yield the same flash point test results as the liquid itself. However, whereas the liquid alone will ignite at that flash point temperature, the solid/liquid mixture will not. *Therefore, flash point tests apply only to liquids.*

If you are uncertain of a waste's physical state, use the *Paint Filter Liquids Test*, page 26.

Spontaneously Combustible Solids

§261.21(a)(2)

Spontaneously combustible solids are wastes that are not liquid and behave as spontaneously combustible at 68 °F and 1 atm pressure. This behavior is described as capable of:

> "causing fire through friction, absorption of moisture or spontaneous chemical changes and, when ignited, burns so vigorously and persistently that it creates a hazard."

Ignitable Compressed Gases

§261.21(a)(3)

The term "compressed gas" and what makes a compressed gas "ignitable" are defined in the regulations.

A "compressed gas" is any containerized material or mixture *that has:*

> ➤ an absolute pressure exceeding 40 psi at 70 °F, or regardless of the pressure at 70 °F, has an absolute pressure exceeding 104 psi at 130 °F, *or*

> ➤ any liquid flammable material having a vapor pressure exceeding 40 psi absolute at 100 °F as determined by ASTM Test D–323.

A compressed gas is "ignitable" if:

> "a mixture of 13% or less (by volume) with air forms a flammable mixture or the flammable range with air is wider than 12% regardless of the lower limit. These limits shall be determined at atmospheric temperature and pressure. The method of sampling and test procedure shall be acceptable to the Bureau of Explosives and approved by the director, Pipeline and Hazardous Materials Technology, U.S. Department of Transportation."

There are three tests established by the Bureau of Explosives as follows:

1. **Flame Projection Apparatus**

    "The flame projects more than 18 inches beyond the ignition source with valve opened fully, or, the flame flashes back and burns at the valve with any degree of valve opening."

2. **Open Drum Apparatus**

"There is any significant propagation of flame away from the ignition source."

3. **Closed Drum Apparatus**

"There is any explosion of the vapor-air mixture in the drum."

Oxidizers

§261.21(a)(4)   Oxidizers are wastes that meet the "oxidizer" definition of 261.21(a)(4):

"a substance such as a chlorate, permanganate, inorganic peroxide, or a nitrate, that yields oxygen readily to stimulate the combustion of organic matter."

The DOT regulatory definition of an oxidizer was contained in §173.151 of 49 CFR, and the definition of an organic peroxide was contained in paragraph 173.151a. *An organic peroxide is a type of oxidizer.* An organic compound containing the bivalent -O-O- structure and which may be considered a derivative of hydrogen peroxide where one or more of the hydrogen atoms have been replaced by organic radicals *must be classed as an organic peroxide unless*:

➤ the material meets the definition of a Class A explosive or a Class B explosive, as defined in §261.23(a)(8), in which case it must be classed as an explosive; *or*

➤ the material is forbidden to be offered for transportation according to 49 CFR 172.101 and 49 CFR 173.21; *or*

➤ it is determined that the predominant hazard of the material containing an organic peroxide is other than that of an organic peroxide; *or*

➤ according to data on file with the Pipeline and Hazardous Materials Safety Administration in the DOT, it has been determined that the material does not present a hazard in transportation.

Wastes that exhibit the ignitability characteristic are assigned EPA Hazardous Waste Number D001.

2. CORROSIVITY

§261.22   The characteristic of corrosivity applies only to aqueous and liquid wastes. A waste is corrosive *if it is:*

➤ aqueous and its pH is less than or equal to 2 or greater than or equal to 12.5, *or*

➤ liquid and it corrodes steel at a rate greater than 6.35 mm per year.

This characteristic is usually the easiest to determine, since it is well-defined and is based on tests that are easy to perform. For aqueous wastes, measure pH of the waste "as is" (without adding water) using a standard pH meter. The steel corrosion test is a standard test of National Association of Corrosion Engineers, TM-01-69.

Keep in mind that this characteristic applies only to aqueous and liquid wastes: if the waste is not amenable to pH measurement and it does not contain liquids as determined by the *Paint Filter Liquids Test*, page 26, it cannot exhibit corrosivity.

Wastes that exhibit the corrosivity characteristic are assigned EPA Hazardous Waste Number D002.

3. REACTIVITY

§261.23

Reactivity is the most nebulous of the characteristics because there are no absolute quantitative tests to define it. This characteristic is meant to identify those wastes that have the potential to explode or release toxic gases during the waste management process. The intent is twofold: to protect the people working with the waste, and to prevent fires and explosions.

When determining whether a waste exhibits reactivity, you need to understand how the waste behaves, and then you need to decide if that behavior fits the regulatory intent for classifying reactive wastes. The combination of your knowledge of the waste, the regulatory description of reactivity (40 CFR 261.23), and the guidelines given here should be enough for you to make a judgment about the reactivity of a particular waste.

The characteristic of reactivity evaluates two different types of hazards: (1) *physical hazards* such as explosions, violent reactions, and fire hazards, and (2) *health hazards* from the release of toxic gases from the waste.

Physical Hazards

The regulatory language describes reactive wastes that present a physical hazard as *reactive wastes that:*

➤ are normally unstable and readily undergo violent change without detonating, *or*

➤ react violently with water, *or*

➤ form potentially explosive mixtures with water, *or*

> ➤ are capable of detonation or explosive reaction if subjected to a strong initiating source or heated under confinement, *or*

> ➤ are readily capable of detonation or explosive decomposition or reaction at standard temperature and pressure, *or*

> ➤ meet the DOT definition of "forbidden explosive" (49 CFR 173.51), "class A explosive" (49 CFR 173.53), or "class B explosive" (49 CFR 173.88).

The preamble to the RCRA regulations of May 19, 1980 states that wastes reactive because of physical hazard potential are those that "because of extreme instability and tendency to react violently or explode, pose a problem at all stages of the waste management process." This statement is useful in determining the degree of physical hazard that is necessary for a waste to be RCRA reactive.

When evaluating a waste for reactivity due to physical hazard potential, compare your waste to these materials, which U.S. EPA has listed as reactive when discarded as commercial chemicals: nitroglycerine, hydrazine, and methyl ethyl ketone peroxide.

Health Hazards

Reactive wastes that pose health hazards are more difficult to identify. The regulations offer this description of a reactive waste that poses a health hazard:

> When exposed to water (or when exposed to pH conditions between 2 and 12.5 if the waste contains cyanide or sulfide) it "generates toxic gases, vapors, or fumes in a quantity sufficient to present a danger to human health or the environment."

This standard is sometimes difficult to assess since various management scenarios will present different levels of hazard. For example, a waste stored in a small, unventilated room may pose a greater hazard to a nearby worker than if the same waste and worker were outside.

In 1985, U.S. EPA issued the following *guidance* for cyanide- and sulfide-bearing wastes for determining whether a waste poses a health hazard sufficient to deem it RCRA reactive, depending on the concentration of *releasible* cyanide or sulfide:

> If a waste contains less than 250 mg/kg total cyanide or 500 mg/kg total sulfide, it is not reactive. If the waste contains over those levels, you need to determine how much of the cyanide or sulfide is "available for release." Wastes are considered reactive which contain more than 250 mg/kg cyanide or 500 mg/kg sulfide which is "available for release."

Although EPA Headquarters revoked this *guidance* in 1998, some states and EPA Regions still use it.

EPA has also stated that a waste is RCRA reactive if OSHA limits for cyanide or sulfide have been exceeded in areas where the waste is generated, stored, or otherwise handled. *Keep in mind that these are EPA guidelines and not part of the regulation.*

The characteristic of reactivity is complex and difficult to determine. A waste may be reactive and yet not reactive to the degree that it exhibits the reactivity characteristic.

Wastes that exhibit the RCRA characteristic of reactivity are assigned EPA Hazardous Waste Number D003.

4. TOXICITY CHARACTERISTIC

§261.24

The toxicity characteristic (TC) is meant to identify those wastes that if disposed of in the environment have the potential of leaching any of 39 constituents in levels at or above regulatory thresholds. These constituents include 8 heavy metals, 4 insecticides, 2 herbicides, and 25 other organic compounds.

The laboratory test for evaluating wastes under the TC is the Toxicity Characteristic Leaching Procedure (TCLP). The TCLP (described in Method 1311, EPA Publication SW-846), extracts an "artificial leachate" from the waste using deionized water and dilute, buffered acetic acid as leaching agents. (Note: If the waste is a liquid with less than 0.5% solids, the extraction step is skipped and the waste itself, after filtering, is analyzed as the extract.) This extract is considered to be an approximation of the worst case leachate that the waste will generate if disposed in what EPA considers a worst case disposal scenario: a municipal landfill. The TCLP uses an acidic leach with acetic acid because municipal landfills tend to be acidic from decomposition of organic matter within the landfill into organic acids including acetic acid.

If a waste's TCLP extract contains any of the constituents at or above the set MCL, the waste exhibits the TC. The MCL for each constituent is set at 100 times a health-based standard (such as a drinking water standard) established for that constituent.

Wastes that exhibit the TC are assigned EPA Hazardous Waste Numbers according to the MCLs exceeded.

TABLE 2-4, page 22, shows TC criteria as follows: EPA Hazardous Waste Numbers, corresponding constituents, and MCLs in mg/l.

Table 2-4. Toxicity Characteristic Criteria

EPA Hazardous Waste Number	Constituent	MCL in Leachate[A] (mg/l)	Calculated Maximum[B] Concentration in Solid (mg/kg)
METALS			
D004	Arsenic	5.0	100
D005	Barium	100.0	2,000
D006	Cadmium	1.0	20
D007	Chromium (total)	5.0	100
D008	Lead	5.0	100
D009	Mercury	0.2	4
D010	Selenium	1.0	20
D011	Silver	5.0	100
INSECTICIDES			
D012	Endrin	0.02	0.4
D013	Lindane	0.4	8
D014	Methoxychlor	10.0	200
D015	Toxaphene	0.5	10
HERBICIDES			
D016	2,4-D	10.0	200
D017	2,4,5-TP (Silvex)	1.0	20
OTHER ORGANICS			
D018	Benzene	0.5	10
D019	Carbon tetrachloride	0.5	10
D020	Chlordane	0.03	0.6
D021	Chlorobenzene	100.0	2,000
D022	Chloroform	6.0	120
D023	o-Cresol	200.0[C]	4,000
D024	m-Cresol	200.0[C]	4,000
D025	p-Cresol	200.0[C]	4,000
D026	Cresols	200.0[C]	4,000

Table 2-4. (*continued*)

EPA Hazardous Waste Number	Constituent	MCL in Leachate[A] (mg/l)	Calculated Maximum[B] Concentration in Solid (mg/kg)
OTHER ORGANICS (*CONTINUED*)			
D027	1,4-Dichlorobenzene	7.5	150
D028	1,2-Dichloroethane	0.5	10
D029	1,1-Dichloroethylene	0.7	14
D030	2,4-Dinitrotoluene	0.13[D]	2.6
D031	Heptachlor (& epoxide)	0.008	0.16
D032	Hexachlorobenzene	0.13[D]	2.6
D033	Hexachlorobutadiene	0.5	10
D034	Hexachloroethane	3.0	60
D035	Methyl ethyl ketone	200.0	4,000
D036	Nitrobenzene	2.0	40
D037	Pentachlorophenol	100.0	2,000
D038	Pyridine	5.0[D]	100
D039	Tetrachloroethylene	0.7	14
D040	Trichloroethylene	0.5	10
D041	2,4,5-Trichlorophenol	400.0	8,000
D042	2,4,6-Trichlorophenol	2.0	40
D043	Vinyl chloride	0.2	4

A. The MCL for each constituent is set at 100 times a health-based standard (such as a drinking water standard) established for that constituent. MCLs in mg/l.

B. See *Using Total Constituent Data for Determining TC*, page 24, for explanation of this column.

C. If cresol isomers cannot be differentiated, total cresols is used.

D. Quantitation limit is greater than the calculated regulatory level. The quantitation limit therefore becomes the regulatory level.

TC Exemptions

There are two exemptions for TC wastes D018 through D043, and there is a third exemption available through a petition process for certain wastes that are hazardous only because they exhibit the TC for chromium.

§261.8 1. Wastes that contain PCBs that (1) are or contain dielectric fluids, and (2) are regulated under TSCA as PCB waste.

§261.4(b)(10) 2. Wastes that are contaminated media and debris from corrective action work required by the underground storage tank regulations of 40 CFR 280.

§261.4(b)(6)(i) 3. On a case-by-case basis, waste that exhibits the TC for chromium may be exempted *if*:

> ➤ the waste is not listed and does not exhibit any other characteristic (including the TC for any constituent other than chromium), *and*

> ➤ the chromium in the waste is exclusively (or nearly exclusively[3]) trivalent chromium, *and*

> ➤ the process that generates the waste uses exclusively (or nearly exclusively[3]) trivalent chromium, *and*

> ➤ the process that generates the waste does not generate hexavalent chromium, *and*

> ➤ the waste is "typically and frequently" managed in non-oxidizing environments.

The third exemption is not self-implementing: it must be granted by EPA or an authorized state. It may be used to exempt waste from an individual site or an entire industry. EPA made this process available recognizing the hazard differences between trivalent and hexavalent chromium. *Trivalent chromium* is generally considered low in toxicity and immobile in the environment. It is commonly used in products such as cosmetics and vitamin pills. *Hexavalent chromium*, however, is considered a human carcinogen and quite mobile in the environment.

Using Total Constituent Data for Determining TC

The TCLP is an expensive test procedure. Rather than perform the TCLP, you may be able to use total constituent data to evaluate wastes under the TC if either (1) the waste is entirely solid with no liquid fraction, or (2) the waste contains less than 0.5% solids.

1. **Waste Is Entirely Solid with No Liquid Fraction**

    For wastes with no liquid fraction, since the TCLP leaches 100 grams of sample with 2,000 grams of leaching solution, providing a maximum of 20 times dilution of constituents in the sample, a waste containing less than 20 times the MCL for any

---

3. The qualifier "or nearly exclusively" is used because of the difficulty in distinguishing between trivalent and hexavalent chromium at low concentrations.

given constituent cannot exhibit the TC for that constituent because even if 100% of the constituent in the waste leaches out, the 20 times dilution of the extraction fluid would lower the concentration of the constituent below its MCL.

For wastes with no liquid fraction, use TABLE 2-4, page 22, to determine whether you can use total constituent data or if you must perform the TCLP.

For wastes that contain over 20 times the MCL for a constituent, no determination can be made—you must perform the TCLP.

If you plan to use total constituent data for organics, be sure that sampling and sample preparation is done such that loss of organics is minimized.

2. **Waste Contains Less Than 0.5% Solids**

For wastes containing less than 0.5% solids, the waste itself may be analyzed as the TCLP extract. However, analyzing the waste itself may be conservative, particularly for metals, since any suspended solids are digested and reported as "total."

The TCLP filters out solids before analysis; therefore, if you have a liquid waste with less than 0.5% solids that exceeds an MCL, you may wish to retest and have the waste filtered per TCLP before analysis.

Hazardous Wastes Classification by EPA Hazardous Waste Numbers

Under RCRA, hazardous wastes are classified by their EPA Hazardous Waste Numbers. An important aspect of hazardous waste classification is compliance with the LDRs since classification determines what treatment standards apply to the waste prior to land disposal.

Three general rules for classifying RCRA hazardous wastes are as follows:

§261.20(b)   1. A non-listed hazardous waste that exhibits more than one characteristic carries EPA Hazardous Waste Numbers for all the characteristics it exhibits.

For example, a waste acid that exhibits corrosivity and exhibits the TC for lead, is named D002/D008.

§268.9(a)   2. A listed hazardous waste that has been listed only as toxic (T) or acute hazardous (H) and also exhibits one or more characteristics carries EPA Hazardous Waste Numbers of the listing and those characteristics it exhibits.

For example, a spent solvent mixture F001 that also exhibits ignitability is named F001/D001.

§268.9(a)

3. A listed hazardous waste *does not* carry the EPA Hazardous Waste Numbers for characteristics or constituents it was listed for.

For example, according to §261 Appendix VII, F006 is listed for containing chromium, cadmium, nickel, and complexed cyanides. If an F006 waste exhibits the TC for chromium or cadmium, it is named only F006. However, if it exhibits the TC for lead, it is named F006/D008.

Solid Wastes Testing

§262.11

RCRA regulations require you to determine if your waste is hazardous, but they do not require you to test your waste to make that determination. Generally, testing cannot be used to identify listed hazardous wastes: you must know how the waste was generated. Testing may be used to evaluate a waste for the four characteristics, or you may use available information or your knowledge of the waste. For example, if your raw materials and process are such that your waste could not possibly contain lead, you may not need to test for lead.

Submit for Laboratory Testing

Often you have no choice but to submit your waste for laboratory testing to determine if the waste is hazardous waste. Sampling and analysis protocol for solid wastes are outlined in EPA guidance document SW-846, "Test Methods for Evaluating Solid Wastes." When arranging for testing with a laboratory, specify that the laboratory's test methods must be consistent with SW-846.

§261.4(d)

A sample of hazardous waste sent to a laboratory for characteristic or composition testing is not subject to regulation as hazardous waste if you (1) package the sample so that it won't leak or vaporize, and (2) include the following information with the sample: your name, address, telephone number; the lab's name, address, telephone number; quantity of sample; date of shipment; and description of sample.

Paint Filter Liquids Test

The Paint Filter Liquids Test is used to determine whether a waste contains free liquid. This determination is crucial for two of the four characteristic tests: ignitability and corrosivity.

The Paint Filter Liquids Test is very simple: Place a 100 gram sample of waste into a conical 60 mesh, or 400 micron, paint filter for five minutes. If after five minutes, any material has dropped from the filter, the waste is considered to contain free liquid. This test procedure is listed in SW-846 as Method 9095.

Documentation Requirements

§262.40(c)

RCRA regulations explicitly require you to document the determinations you make about whether a solid waste is a hazardous waste. This determination and documentation are required for each solid waste you generate, regardless of how obvious the determination is to you.

This documentation may include test results, raw materials information, and process data: any information you used to characterize the solid waste and determine if it is hazardous under the regulations.

The *Solid Waste Inventory and Profile Sheet*, page 28, can help you comply with this requirement.

Solid Waste Inventory and Profile Sheet

Waste Common Name

Description

Generating Process

Physical Form

Regulatory Status

☐ EPA Hazardous ☐ State Hazardous ☐ Exempt ☐ Universal Waste ☐ Used Oil

EPA/State Hazardous Waste Numbers

Listings

☐ Non-specific Source Wastes   F ___ ___ ___
☐ Specific Source Wastes   K ___ ___ ___
☐ Discarded Commercial Chemical Products   P ___ ___ ___   U ___ ___ ___

Characteristics		Determined by Testing	Knowledge	Documentation Attached
☐ Ignitable	D001	☐	☐	☐
☐ Corrosive	D002	☐	☐	☐
☐ Reactive	D003	☐	☐	☐
☐ Toxic by TC	D___	☐	☐	☐

Use as Fuel

☐ Hazardous Waste Fuel
☐ Used Oil Fuel   ☐ Specification   ☐ Off-Specification
☐ Burned Onsite   ☐ Burned Offsite   ☐ Marketed Directly to Burner

Exemptions Claimed

☐ Solid Waste Exclusion: §261.4(a) (___)   ☐ Delisted
☐ Hazardous Waste Exclusion: §261.4(b) (___)   ☐ Other: _____
☐ Reuse, Recycling Exemption (§261.1, §261.2)   ☐ Other: _____

Comments

Figure 2-1. Solid Waste Inventory and Profile Sheet

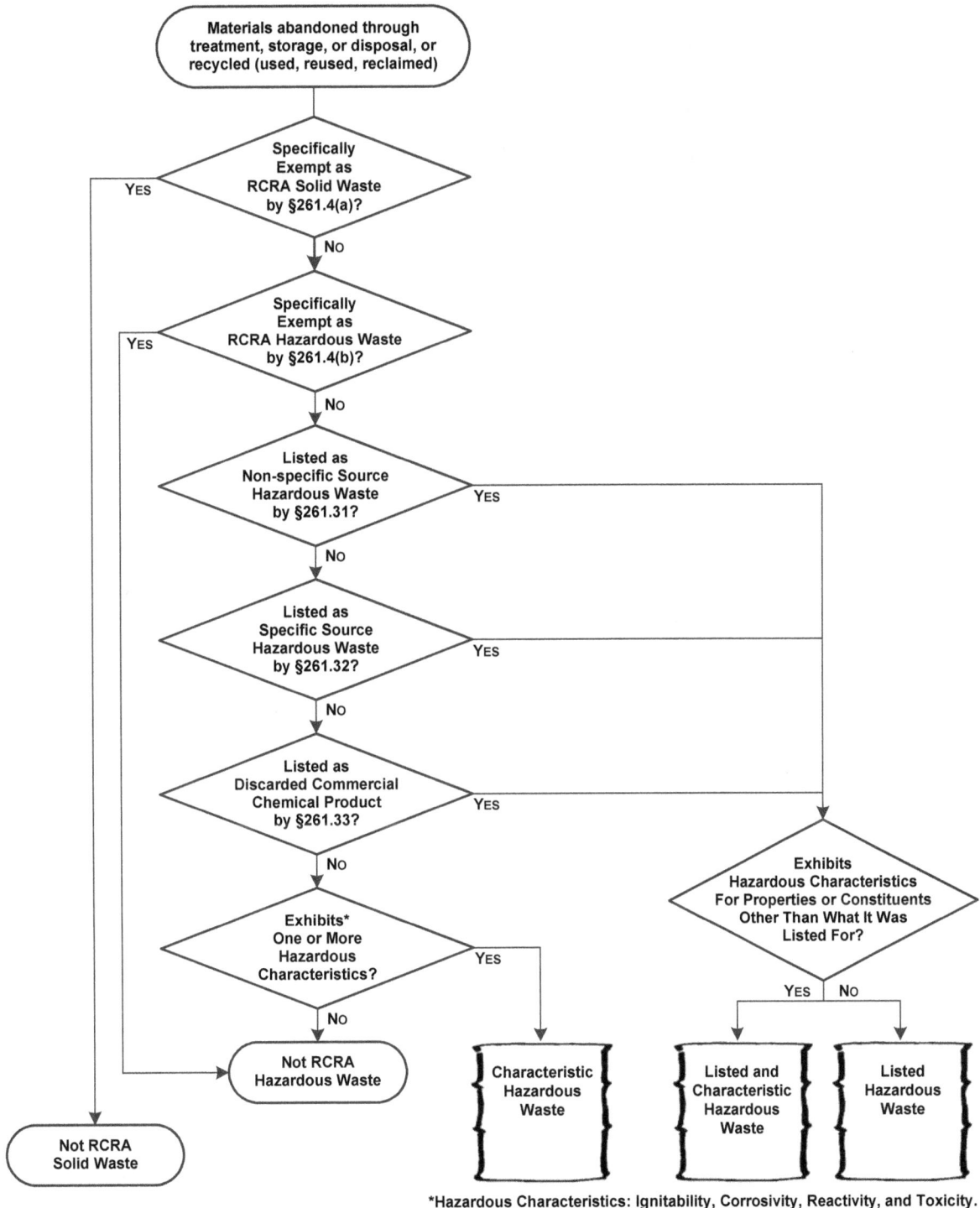

Figure 2-2. Simplified Look at Hazardous Waste Definition

CHAPTER 2: REVIEW QUESTIONS

Questions

1. Which of the following is a RCRA waste list?
    (Check all that apply.)

    ☐ A. Non-specific Source Wastes list
    ☐ B. Specific Source Wastes list
    ☐ C. Discarded Commercial Chemical Products list
    ☐ D. IARC Carcinogen list
    ☐ E. Hazardous Substance list

2. Waste that contains no detectable levels of hazardous constituents and exhibits no hazardous characteristics is not a RCRA hazardous waste.

    ☐ Truth or ☐ Myth?

3. Which of the following would be regulated as a non-specific source hazardous waste?
    (Check all that apply.)

    ☐ A. waste paint with 20% toluene
    ☐ B. spent mineral spirits used for cleaning
    ☐ C. spent degreaser, 15% trichloroethylene (TCE) prior to use

4. Which of the following would be F003?
    (Check all that apply.)

    ☐ A. unused xylene
    ☐ B. xylene used for paint clean up
    ☐ C. spill residue of xylene
    ☐ D. spent solvent from using 20% xylene, 80% mineral spirits

5. Which of the following would be regulated as a P-listed or U-listed waste? (Check all that apply.)

   ☐ A. spill residue containing spent TCE
   ☐ B. spill residue containing virgin TCE
   ☐ C. groundwater that contains virgin TCE
   ☐ D. old pesticide to be discarded

6. Which of the following is a RCRA hazardous characteristic? (Check all that apply.)

   ☐ A. ignitability
   ☐ B. corrosivity
   ☐ C. explosivity
   ☐ D. reactivity
   ☐ E. toxicity

7. Which of the following wastes might exhibit ignitability? (Check all that apply.)

   ☐ A. spent mineral spirits, flash point 102 °F
   ☐ B. DOT combustible liquid
   ☐ C. rags that are spontaneously combustible
   ☐ D. wood soaked with gasoline

8. Which of the following wastes might exhibit corrosivity? (Check all that apply.)

   ☐ A. sodium hydroxide pellets
   ☐ B. spent caustic, pH 12.7
   ☐ C. very salty liquid
   ☐ D. aqueous gel, pH 1.7

9. Which of the following describes RCRA reactive wastes? (Check all that apply.)

   ☐ A. reacts violently with water

☐ B. normally unstable

☐ C. forms potentially explosive mixtures with water

☐ D. generates hydrogen cyanide when mixed with hazardous waste acids

10. Which of the following are TC constituents?
    (Check all that apply.)

☐ A. all chemicals with LD50 <50 mg/kg

☐ B. all chemicals that have a primary drinking water standard

☐ C. 8 metals, 4 pesticides, 2 herbicides, and 25 other organics

☐ D. any RCRA hazardous constituent

11. How is D007 pronounced?

☐ A. Doo-Seven

☐ B. Dee-Zero-Zero-Seven

☐ C. Doo-Oh-Seven

☐ D. Dee-Double-Oh-Seven

12. You must perform the TCLP to determine if a waste exhibits the TC.

☐ Truth or ☐ Myth?

13. If I do not generate hazardous waste, I am not subject to any of RCRA's hazardous waste regulations.

☐ Truth or ☐ Myth?

Answers

1. ☑ A. The Non-specific Source Wastes list is a RCRA list. The *Non-specific Source Wastes* list (F list) is the "generic" list of hazardous wastes. Each listing defines multiple wastes that are generated by various industries and processes.

   ☑ B. The Specific Source Wastes list is a RCRA list. The *Specific Source Wastes* list (K list) defines hazardous wastes by industry. Each listing is specific by industry and process and usually defines one particular waste.

   ☑ C. The Discarded Commercial Chemical Products list is a RCRA list. The *Discarded Commercial Chemical Products* list (P and U lists) are those chemicals that are hazardous waste if discarded.

   ☒ D. The IARC Carcinogen list is not a RCRA list.

   ☒ E. The Hazardous Substance list is not a RCRA list.

2. ☑ Myth

   Waste that contains no detectable levels of hazardous constituents and exhibits no hazardous characteristics is not a RCRA hazardous waste is a myth. Wastes are regulated as listed hazardous wastes based solely upon meeting the listing description, regardless of their chemical makeup, characteristics, or hazards.

3. ☒ A. Waste paint with 20% toluene would not be regulated as a non-specific source hazardous waste. Toluene is present as an ingredient in the paint, not as a spent solvent.

   ☒ B. Spent mineral spirits used for cleaning would not be regulated as a non-specific source hazardous waste. Mineral spirits is not listed.

   ☑ C. Spent degreaser, 15% TCE prior to use would be regulated as a non-specific source hazardous waste. TCE is a listed solvent under 40 CFR 261.31 and was present in the solvent prior to use to render the spent solvent a listed hazardous waste.

4. ☒ A. Unused xylene would not be F003. Unused xylene would be classified as discarded commercial chemical product U239 if discarded.

   ☑ B. Xylene used for paint clean up would be F003.

   ☒ C. Spill residue of xylene would not be F003. Spill residue of xylene would be classified as a discarded commercial chemical product U239 if discarded.

   ☒ D. Spent solvent from using 20% xylene, 80% mineral spirits would not be F003 because it is not solely F003 listed chemicals. (It may be ignitable if its flash point is <140 °F.)

5. ☒ A. Spill residue containing spent TCE would not be regulated as a P-listed or U-listed waste since the TCE is spent. It would be F-listed.

    ☑ B. Spill residue containing virgin TCE would be regulated as a P-listed or U-listed waste. Spill residue containing virgin TCE could be classified as discarded commercial chemical product.

    ☑ C. Groundwater that contains virgin TCE would be regulated as a P-listed or U-listed waste. Groundwater that contains virgin TCE could be classified as discarded commercial chemical product.

    ☑ D. Old pesticide to be discarded could be regulated as a P-listed or U-listed waste if it is listed.

6. ☑ A. Ignitability is a RCRA hazardous characteristic under 40 CFR 261.21.

    ☑ B. Corrosivity is a RCRA hazardous characteristic under 40 CFR 261.22.

    ☒ C. Explosivity is not a RCRA hazardous characteristic. There is no "explosivity" characteristic. Wastes that have explosive properties are evaluated under the reactivity characteristic.

    ☑ D. Reactivity is a RCRA hazardous characteristic under 40 CFR 261.23.

    ☑ E. Toxicity is a RCRA hazardous characteristic under 40 CFR 261.24.

7. ☑ A. Spent mineral spirits, flash point 102 °F would exhibit ignitability. Spent mineral spirits is ignitable because its flash point is below 140 °F.

    ☒ B. DOT combustible liquid would not exhibit ignitability. It is not ignitable since as a DOT "combustible liquid" its flash point would be above 140 °F.

    ☑ C. Rags that are spontaneously combustible might exhibit ignitability. Rags that exhibit spontaneous combustible behavior need to be further evaluated to determine whether they are ignitable due to spontaneous combustion as defined under 40 CFR 261.21.

    ☒ D. Wood soaked with gasoline would not likely exhibit ignitability. Wood soaked with gasoline that contains no free liquids cannot be flash point tested.

8. ☒ A. Sodium hydroxide pellets would not exhibit corrosivity because in pellet form it is not aqueous; therefore, its pH cannot be measured.

    ☑ B. Spent caustic, pH 12.7 might exhibit corrosivity because its pH is above the regulatory trigger of 12.5 pH units.

    ☑ C. Very salty liquid might exhibit corrosivity. Very salty liquid would be corrosive if it fails the steel corrosion test.

☑ D. Aqueous gel, pH 1.7 might exhibit corrosivity because its pH is below the regulatory trigger of 2.0 pH units.

9. ☑ A. Reacts violently with water describes RCRA reactive wastes.

   ☑ B. Normally unstable describes RCRA reactive wastes.

   ☑ C. Forms potentially explosive mixtures with water describes RCRA reactive wastes.

   ☑ D. Generates hydrogen cyanide when mixed with hazardous waste acids describes RCRA reactive wastes.

10. ☒ A. All chemicals with LD50 <50 mg/kg are TC constituents is an incorrect statement.

   ☒ B. All chemicals that have a primary drinking water standard are TC constituents is an incorrect statement.

   ☑ C. Eight metals, 4 pesticides, 2 herbicides, and 25 other organics are TC constituents. The chemicals that can cause a waste to exhibit the TC are quite limited.

   ☒ D. Any RCRA hazardous constituent are TC constituents is an incorrect statement.

11. ☒ A. Some people pronounce it as Doo-Seven, but it is not the preferred pronunciation.

   ☑ B. Dee-Zero-Zero-Seven is the widely accepted pronunciation.

   ☒ C. Doo-Oh-Seven is not the accepted pronunciation.

   ☒ D. Dee-Double-Oh-Seven is not the accepted pronunciation.

12. ☑ Myth

You must perform the TCLP to determine if a waste exhibits the TC is a myth. You are never required to test a waste to determine if it meets a characteristic. You may use generator knowledge. Sometimes you have enough information about a waste's composition that you do not need to test it.

13. ☑ Myth

If I do not generate hazardous waste, I am not subject to any of RCRA's hazardous waste regulations is a myth. 40 CFR 262.11 requires every solid waste to be evaluated to determine whether it is a hazardous waste, and 40 CFR 262.40(c) requires recordkeeping of that determination.

CHAPTER 3:
Used Oil

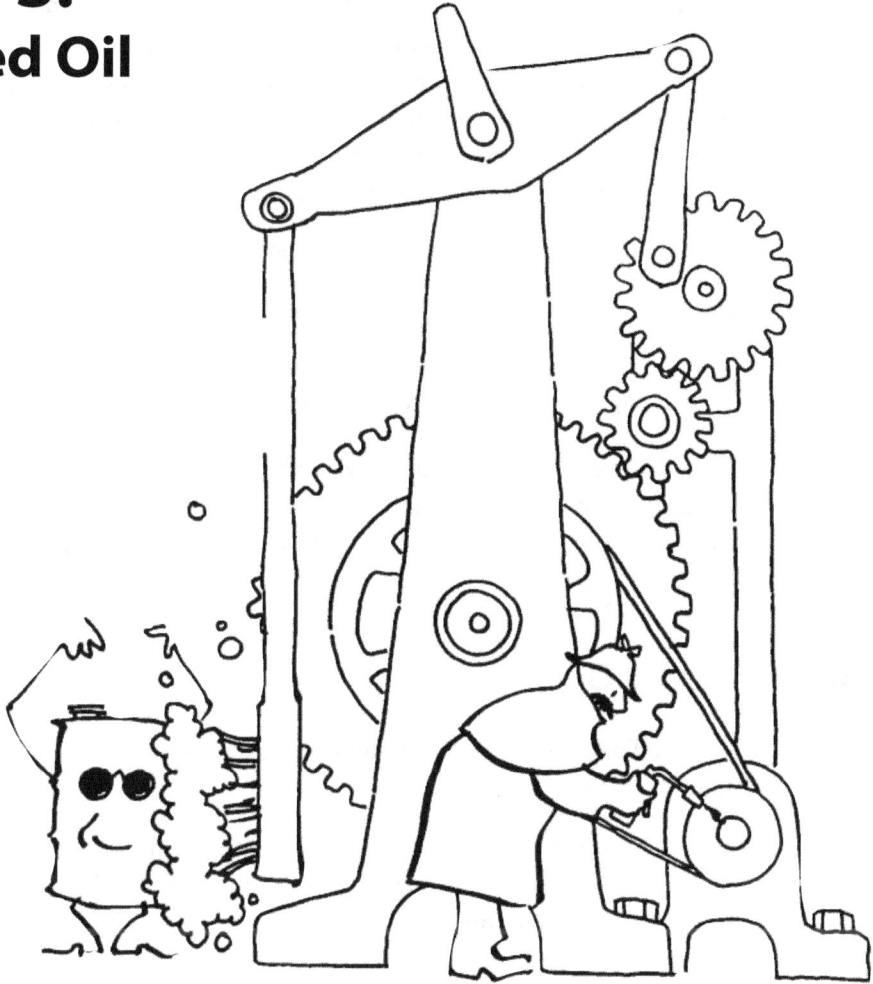

S O PERVASIVE IS HUMAN use of oil that if somehow by magic all of the oil currently in use suddenly disappeared, routine human activity would abruptly halt, and we would find ourselves in a dark, surreal world littered with smoldering, broken, melted machines. Hydraulics, lubrication, heat transfer, and electrical systems all depend on the wonderful properties of oil, making oil a favorite beverage of our modern technological culture.

However, the wonderful properties of an oil usually diminish by use and time, reducing its effectiveness: additives deplete, oil chemistry breaks down, water is absorbed, acids form, and contaminants build up. After the oil becomes used, RCRA regulates what happens to it next.

The Used Oil Recycling Act of 1980 amended RCRA and gave EPA authority to regulate used oil. The history of EPA's used oil regulations development is long and twisted, filled with congressional impatience, lawsuits, federal court admonishment, and tearful pleas and testimony from gas station owners across the country. Out of all this, EPA issued rules that address three environmental concerns: (1) disposal of hazardous wastes into used oil, (2) air emissions from burning used oil as fuel, and (3) spills and leaks of used oil into the environment.

This chapter provides a series of six questions that are applicable as follows:

- Answer questions 1–3 to classify a used oil under RCRA.
- Answer questions 1–5 if you burn used oil or send it directly to a burner.
- Answer questions 1–6 if you claim that used oil is "specification used oil fuel."

Once you have answered the questions that apply to your situation, read through the discussion of the regulations that apply to your role in regard to used oil. The final discussion is of miscellaneous regulations about prohibitions, exemptions, and PCBs.

Question 1: Is It Used Oil?

§279.1 **Is it used oil?** This is a seemingly obvious but necessary question. RCRA's definition of used oil is as follows:

> "Any oil that has been refined from crude oil, or any synthetic oil, that has been used and as a result of such use is contaminated by physical or chemical impurities."

This definition gives three criteria for a material to be "used oil."

1. It needs to be either refined crude oil or synthetic oil—that eliminates animal and vegetable oils.

2. It needs to be "used." In RCRA this means "used as an oil," such as lubrication, hydraulics, and heat transfer. For example, a light end petroleum fraction such as mineral spirits that has been used as a solvent cannot be a "used oil."

   Also, at the risk of stating the obvious, unused oils, for example oily sludge from virgin oil tanks and cleanup from virgin oil spills, are outside the scope of the used oil regulations.

3. "Contaminated by physical or chemical impurities" is taken at its plain meaning.

Question 2: Is the Used Oil Mixed with Hazardous Waste?

§279.10(b)   **Is the used oil mixed with hazardous waste?** This gets tricky. The mixture may be regulated as hazardous waste or used oil depending on the type of hazardous waste and whether the resulting mixture exhibits any hazardous characteristics.

If you have a mixture of a used oil and hazardous waste, work through the flowchart in FIGURE 3-1, page 45, to determine its regulatory status.

Question 3: Does the Used Oil Contain More Than 1,000 ppm Total Halogen?

§279.10(b)(ii)   **Does the used oil contain more than 1,000 ppm total halogen?** RCRA calls this test the "rebuttable presumption." If a used oil contains more than 1,000 ppm total halogen, it is presumed that spent chlorinated solvent has been mixed into the used oil, unless you can prove otherwise.

If you cannot rebut the presumption by showing that the oil has not been adulterated by mixing with any of the halogenated spent solvents listed in F001 and F002, you must classify and manage the oil as hazardous waste. EPA recommends ASTM method D808-81 (total chlorine) for total halogen determination.

This "rebuttable presumption" test is used only to evaluate used oils—it has no other application for determining hazardousness. The rebuttable presumption test does not apply to (1) certain metalworking oils that contain chlorinated paraffins and are reclaimed through tolling arrangements, and (2) used oils from refrigeration or air conditioning systems contaminated with CFCs that are going to be reclaimed.

Question 4: Is the Used Oil "Used Oil Fuel"?

Is the used oil "used oil fuel"? "Used oil fuel" is used oil that (1) has not mixed with listed hazardous waste, (2) does not exhibit a characteristic from being mixed with hazardous waste, and (3) is burned for energy recovery or is used to produce a fuel. Used oil fuel may exhibit a hazardous characteristic if it was mixed with a waste hazardous solely for ignitability or if it exhibits a characteristic on its own.

Mixtures are tricky—work through the flowchart in FIGURE 3-1, page 45.

Question 5: Is the Used Oil "Hazardous Waste Fuel"?

§279.10

Is the used oil "hazardous waste fuel"? Used oil is "hazardous waste fuel" if it either (1) is hazardous from being mixed with hazardous waste, or (2) contains over 1,000 ppm total halogen and the presumption of hazardousness cannot be rebutted. Regulations that apply to hazardous waste fuels are found in 40 CFR 266 Subpart H.

Mixtures are tricky—work through the flowchart in FIGURE 3-1, page 45.

Question 6: Is the Used Oil Fuel "Specification Used Oil Fuel"?

§279.11

Is the used oil fuel "specification used oil fuel"? Except for the two recordkeeping requirements noted below, "specification used oil fuel" is exempt from RCRA used oil regulations. The burning of it is not restricted. In TABLE 3-1, note that the constituent levels are not leachate results, but the concentrations of constituents in the oil.

Table 3-1. Specification Used Oil Fuel

Constituent/Property	Allowable Level
Arsenic	≤ 5 ppm
Cadmium	≤ 2 ppm
Chromium	≤ 10 ppm
Lead	≤ 100 ppm
Flash Point	≥ 100 °F
Total Halogens	≤ 4,000 ppm

RECORDKEEPING REQUIREMENTS

If you claim that a used oil is specification used oil fuel, you must maintain the following records:

§279.72

1. Analyses or other documentation showing that the used oil meets the specification.

§279.74(b) 2. Records of each shipment showing where the oil was shipped, the quantity, dates of shipment, and a cross-reference to the documentation maintained under item 1.

The philosophy behind the specification and exemption is that burning used oil that meets the specification does not present any greater hazard than burning virgin oil. If your used oil fuel does not meet the specification, you may blend it so that it does meet the specification if you comply with the regulations for processors and re-refiners (see Processors/Re-refiners, page 42).

At first thought, the specifications, most notably the flash point and halogen limits, seem inconsistent with the criteria for determining if a used oil is a hazardous waste. However, by design the regulations differentiate between used oils that have been intentionally mixed with hazardous waste and those that have not.

Generators

§279 Subpart C Generators must determine if their used oil is *hazardous waste* or *hazardous waste fuel*, and manage those materials as such.

Generators must store their used oil in tanks or containers that are in good condition and are labeled "used oil." Releases of used oil must be cleaned up. When it comes time to ship used oil, generators must use a transporter who has an EPA ID number.

The following two situations do not require a transporter who has an EPA ID number:

1. Used oil reclaimed under a tolling arrangement contract. The contract must indicate the type of used oil and frequency of shipments, that the vehicle is owned and operated by the reclaimer, and that the reclaimed oil will be returned to the generator.

2. A generator may transport "small amounts" (less than 55 gallons at a time) of used oil to either a "used oil collection center" or a "used oil aggregation point" as defined below.

Used Oil Collection Centers

§279 Subpart D Sites that collect and store used oil only from household "do-it-yourself" (DIY) oil changes must comply with the generator standards. "Used oil collection centers" are sites that accept small (55 gallons or less) shipments from generators. They may also accept DIY used oil. In addition to complying with the generator standards, a used oil collection center must be registered/licensed/permitted or otherwise recognized as a used oil management facility by a state or local authority.

Used Oil Aggregation Points

§279 Subpart D   A "used oil aggregation point" collects and stores small (55 gallons or less) used oil shipments only from generating sites owned or operated by the same entity that owns or operates the used oil aggregation point. It may also accept DIY used oil. Used oil aggregation points must comply with the generator standards.

Transporters/Transfer Facilities

§279 Subpart E   Transporters must have an EPA ID number. A truck that has been used for transporting hazardous waste must be decontaminated before transporting used oil. Transporters must determine the halogen content of used oil they transport. Used oil shipments must be delivered to another transporter, a recycling facility, or a used oil burner. Transporters are also required to respond to releases during transportation and maintain tracking records of used oil pickups and deliveries.

Transfer facilities that are used to store used oil shipments for 35 days or less must store their used oil in tanks or containers that are in good condition and are labeled "used oil." Releases of used oil must be cleaned up. The facility must have secondary containment consisting of a floor and walls, dikes, or berms that are "sufficiently impervious" to oil. Transfer facilities that store for longer than 35 days are subject to the same standards as processors/re-refiners.

Processors/Re-refiners

§279 Subpart F   Standards for processors/re-refiners are more stringent than those for other used oil management activities. In addition to having an EPA ID number; determining the halogen content of used oil managed; storing in tanks or containers that are labeled, in good condition, and have secondary containment; and responding to used oil releases, there are requirements that mimic the hazardous waste regulations. The facility must have emergency and contingency plans, and it must comply with preparedness and prevention requirements. No written closure plan is required; however, there are closure performance standards for facilities that close, including post-closure requirements as a hazardous waste landfill if soils cannot be decontaminated. The facility must have a written sampling and analysis plan specifying how it complies with the rebuttable presumption and the used oil specification. The facility must also maintain tracking records of incoming used oil and outgoing recycled products and submit a biennial report to EPA or authorized state.

Burners

§279 Subpart G   Burners (those who burn off-specification used oil fuel) must notify EPA of burner status; have an EPA ID number; determine the halogen content of used oil burned; and store in tanks or containers that are labeled, in good condition, and have secondary containment. Releases of used oil must be cleaned up. Burners must maintain records of each shipment accepted. Before accepting a shipment, burners must send a one-time notice to whomever sent the oil to them, certifying that the burner has notified EPA of its used oil burning and will burn the oil only in an industrial furnace or boiler as defined in §260.10. Burners must maintain copies of these notices.

Marketers

§279 Subpart H   A marketer of used oil fuel is someone who ships off-specification used oil fuel directly to a burner, or *someone other than a generator or transporter who receives oil from a generator* who first claims that a used oil is specification used oil fuel.

A marketer must have an EPA ID number; track and maintain records of used oil fuel delivery to the burner; and ship off-specification used oil fuel only to a burner who has an EPA ID number and will burn the oil only in an industrial furnace or boiler as defined in §260.10. Also, marketers must obtain the one-time notice from burners regarding burner notification and compliance with burning restrictions. A marketer who first claims that a used oil meets the specification must maintain documentation substantiating the claim.

Prohibitions, Exemptions, and PCBs

PROHIBITIONS

§279.12   Placing used oil in surface impoundments or waste piles is prohibited without a RCRA hazardous waste permit. Using used oil for dust suppressant is prohibited, unless done under an approved state program.

EXEMPTIONS

§279.10(c), (f) Materials contaminated with used oil, such as rags, that will not be burned and do not contain visible free-flowing oil are exempt from regulation as used oil. Also exempt is wastewater, subject to regulation under the Clean Water Act, that contains de minimis quantities of used oil.

PCBs

§279.10(i) Used oil fuels containing 2 ppm to 49 ppm PCBs are regulated by TSCA 40 CFR 761.20(e) *and* RCRA used oil regulations. Used oils with 50 ppm or greater PCBs are regulated *only* by TSCA.

A Final Note About Used Oil

Used oil regulations are *slippery*. EPA has tried to balance two interests: satisfying regulatory concerns while encouraging used oil recycling. This balancing act has resulted in a complex set of regulations. Keep in mind that these regulations are offspring of RCRA, and although used oils unadulterated with hazardous waste are not regulated as hazardous waste, many of RCRA's grim enforcement provisions apply to persons who generate and manage used oil. If you know your used oils, how they are managed, what is (and is not) mixed into them, and what happens to them when they leave your plant, you are better able to keep the long, sharp teeth of RCRA enforcement away from your door.

*Specification: Arsenic ≤5 ppm | Cadmium ≤2 ppm | Chromium ≤10 ppm | Lead ≤100 ppm | Flash Point ≥100 °F | Total Halogen ≤4,000 ppm.

Figure 3-1. Simplified Look at Used Oil Characterization

CHAPTER 3: REVIEW QUESTIONS

Questions

1. Which of the following are criteria for oil to be "used oil" under RCRA?
   (Check all that apply.)

   ☐ A. is derived from petroleum
   ☐ B. has been used as an oil
   ☐ C. has been contaminated through use
   ☐ D. contains a RCRA hazardous constituent

2. Used oil containing halogens or metals or other hazardous substance is hazardous waste.

   ☐ Truth or ☐ Myth?

3. The "rebuttable presumption":
   (Check all that apply.)

   ☐ A. only applies to used oils that contain over 1,000 ppm total halogen
   ☐ B. presumes that halogen content is from mixing chlorinated solvents into used oil
   ☐ C. doesn't apply to certain specifically listed used oils
   ☐ D. can only be rebutted through a formal petition process

4. Used cooking oil used to fuel boats could be classified as hazardous waste fuel.

   ☐ Truth or ☐ Myth?

5. No detectable amounts of halogens or metals is allowed in used oil fuels because these substances will cause air pollution.

   ☐ Truth or ☐ Myth?

6. EPA issues ID numbers for anyone who collects used cooking oil to make biodiesel.

   ☐ Truth or ☐ Myth?

7. Used oil fuels that contain PCBs at very low concentrations are regulated by which one of the following?

☐ A. RCRA only

☐ B. RCRA and TSCA

☐ C. RCRA or TSCA depending on the concentration

☐ D. state regulations only

8. Rags contaminated with used oil are regulated as used oil under the RCRA Mixture Rule.

☐ Truth or ☐ Myth?

Answers

1. ☑ A. Oil derived from petroleum is a criterion for RCRA "used oil."

    ☑ B. Oil that has been used as an oil is a criterion for RCRA "used oil."

    ☑ C. Oil that has been contaminated through use is a criterion for RCRA "used oil."

    ☒ D. Oil that contains a RCRA hazardous constituent is a criterion for RCRA "used oil" is an incorrect statement. Oil does not have to contain any specific contaminants to be RCRA "used oil."

2. ☑ Myth

    Used oil containing halogens or metals or other hazardous substance is hazardous waste is a myth. The classification of used oil as hazardous waste depends on (1) whether hazardous waste has been mixed with the used oil, (2) the type of hazardous waste, and (3) whether the used oil mixture exhibits a characteristic.

3. ☑ A. The "rebuttable presumption" only applies to used oils that contain over 1,000 ppm total halogen.

    ☑ B. The "rebuttable presumption" presumes that halogen content is from mixing chlorinated solvents into used oil. Therefore, the presumption can be rebutted by showing that the halogen content is not from mixing spent solvents into the oil.

    ☑ C. The "rebuttable presumption" doesn't apply to certain specifically listed used oils.

    ☒ D. The "rebuttable presumption" can only be rebutted through a formal petition process is an incorrect statement. You should have the documentation on file, but it does not need to be approved by or submitted to anyone.

4. ☑ Myth

    Used cooking oil used to fuel boats could be classified as hazardous waste fuel is a myth. A used oil must be derived from crude oil or be synthetic. Animal or vegetable oils cannot meet the definition of RCRA used oil.

5. ☑ Myth

    No detectable amounts of halogens or metals is allowed in used oil fuels because these substances will cause air pollution is a myth. However, burning used oil fuel that contains greater than "specification used oil fuel" levels is restricted.

6. ☑ **Myth**

   EPA issues ID numbers for anyone who collects used cooking oil to make biodiesel is a myth. Cooking oil would not be refined from crude or synthetic, so it would not be "used oil" under RCRA.

7. ☒ A. Used oil fuels that contain PCBs at very low concentrations are regulated by RCRA only is an incorrect statement.

   ☑ B. Used oil fuels that contain PCBs at very low concentrations are regulated by RCRA and TSCA. Used oil fuel with 2 ppm to 49 ppm PCBs is regulated under both RCRA and TSCA.

   ☒ C. Used oil fuels that contain PCBs at very low concentrations are regulated by RCRA or TSCA depending on the concentration is an incorrect statement.

   ☒ D. Used oil fuels that contain PCBs at very low concentrations are regulated by state regulations only is an incorrect statement.

8. ☑ **Myth**

   Rags contaminated with used oil are regulated as used oil under the RCRA Mixture Rule is a myth. Rags with no visible free-flowing oil that will not be burned are not regulated as used oil. (The RCRA Mixture Rule is discussed in CHAPTER 4: *Other Waste Characterization Principles: Mixture Rule*, page 52.)

CHAPTER 4:
Other Waste Characterization Principles

I N THIS CHAPTER, THE following waste characterization principles are discussed: Empty Container Rule, Mixture Rule, Derived-From Rule, Contained-In Policy for contaminated environmental media, Contained-In Rule for debris, and the characterization of spill residues.

Empty Container Rule

§261.7   A container is considered empty by RCRA definition *if:*

> ➤ all material has been removed from the container that can be removed using the practices that are commonly used (e.g., pouring, pumping), *and*

> ➤ there is no more than 1 inch of residue on the bottom of the container, *or*

> ➤ for containers 119 gallons or less, remaining residue is no more than 3% by weight of the total capacity of the container, *or*

> ➤ for containers greater than 119 gallons, remaining residue is no more than 0.3% by weight of the total capacity of the container.

An empty container is not regulated as RCRA hazardous waste, except that an empty container which has held an acute hazardous material [40 CFR 261.33(e) list] or an acute hazardous waste [EPA Hazard Code (H)] is hazardous waste unless the liner has been removed or the container has been triple-rinsed with a suitable solvent. Other methods of decontamination are acceptable if they are generally recognized as effective in scientific literature. Liners or rinsings from this container decontamination are regulated as hazardous wastes.

Once a container is considered empty under RCRA, residue that is removed from the container is not regulated as hazardous waste unless it exhibits any of the four hazardous characteristics.

The empty container criteria applies to tank trucks, tote tanks, roll-off containers, and anything else that meets the RCRA definition of a "container":

> "Container means any portable device in which a material is stored, transported, treated, disposed of, or otherwise handled."

Mixture Rule

There are very specific rules dealing with mixtures involving hazardous wastes. The most common mixtures are in TABLE 4-1. There are also exemptions for mixtures of wastewaters from petroleum refining and carbamates production.

Table 4-1. Mixture Rule Classification

Mixture Composition			Regulated As	Regulatory Citation
Listed hazardous waste.	+	Solid Waste	Stays regulated as that listed waste, regardless of the quantity of the listed waste present in the mixture.	§261.3 (a)(2)(iv)
Characteristic hazardous waste.	+	Any Material	Hazardous only if the mixture exhibits a hazardous characteristic.	§261.3 (b)(3)
Waste listed as hazardous solely because it exhibits ignitability, corrosivity, or reactivity.	+	Solid Waste	Hazardous only if the mixture exhibits a hazardous characteristic.	§261.3 (g)(2)(i)

Table 4-1. (*continued*)

Mixture Composition			Regulated As	Regulatory Citation
Spent solvents* if the total weekly usage of these solvents divided by the average weekly wastewater flow is *1 ppm or less*. *Benzene, carbon tetrachloride, tetrachloroethylene, and trichloroethylene.	+	Wastewaters	Not hazardous *if:* ➤ the mixture does not exhibit a hazardous characteristic, *and* ➤ the wastewater's ultimate discharge is subject to regulation under the Clean Water Act.	§261.3 (a)(2)(iv) (A)
Spent solvents* if the total weekly usage of these solvents divided by the average weekly wastewater flow is *25 ppm or less*. *Cresols, cresylic acid, nitrobenzene, toluene, methyl ethyl ketone, carbon disulfide, isobutanol, pyridine, chlorofluorocarbons, 2-ethoxyethanol, methylene chloride, 1,1,1-trichloroethane, chlorobenzene, and o-dichlorobenzene.	+	Wastewaters	Not hazardous *if:* ➤ the mixture does not exhibit a hazardous characteristic, *and* ➤ the wastewater's ultimate discharge is subject to regulation under the Clean Water Act.	§261.3 (a)(2)(iv) (B)
Discarded commercial chemical product or intermediate (listed in §261.33) from de minimis losses* from manufacturing operations. *De minimis losses include spills and leaks from normal materials handling, minor leaks from process equipment or containers, pump leaks, sample purgings, discharges from relief devices, safety showers, rinsings from cleaning personal protective equipment, and containers.	+	Wastewaters	Not hazardous *if:* ➤ the mixture does not exhibit a hazardous characteristic, *and* ➤ the wastewater's ultimate discharge is subject to regulation under the Clean Water Act.	§261.3 (a)(2)(iv) (D)
Listed toxic (T) wastes in plant laboratory wastewater *if:* ➤ the annualized average lab wastewater flow is no greater than *1%* of the total plant's wastewater flow, *or* ➤ the wastes are no greater than *1 ppm* in the total wastewater flow.	+	Wastewaters	Not hazardous *if:* ➤ the mixture does not exhibit a hazardous characteristic, *and* ➤ the wastewater's ultimate discharge is subject to regulation under the Clean Water Act.	§261.3 (a)(2)(iv) (E)

Derived-From Rule

§261.3(c)(2)(i)   The Derived-From Rule applies to leachate, sludge, incinerator ash, treatment residues, and spill residues: any solid waste generated from the treatment, storage, or disposal of hazardous waste. Use TABLE 4-2, page 54, to help classify derived-from wastes under RCRA.

Table 4-2. Derived-From Rule Classification

Waste Derived From	Regulated As	Regulatory Citation
A waste listed solely for ignitability, corrosivity, or reactivity.	Hazardous only if it exhibits a hazardous characteristic.	§261.3 (g)(2)(ii)
Any other listed hazardous waste.	Listed waste unless delisted.	§261.3 (d)(2)
Solely from characteristic hazardous waste.	Hazardous only if it exhibits a hazardous characteristic.	§261.3 (d)(1)

Contained-In Policy

The RCRA Mixture Rule and the Derived-From Rule do not apply to environmental media such as soil and groundwater that contain listed hazardous waste because environmental media are not solid wastes.

According to this "Contained-In Policy," an environmental medium—for example, groundwater—contaminated with listed hazardous waste is regulated as hazardous waste only as long as it contains the hazardous waste.

The practical effect is that decontaminated soil or groundwater that had contained listed hazardous waste ceases to be regulated as hazardous waste without having to be delisted. However, the level considered "decontaminated" is not well-defined and must be negotiated case-by-case with the RCRA implementing agency.

Contained-In Rule

§268.2(g)(h)

The Contained-In Rule, which is part of the LDR regulations, applies to "hazardous debris."

A material needs three properties to be debris:

1. *Solid material.* Debris must be solid in the literal sense—retains its shape without a container. *Debris may contain liquids;* for example, concrete with liquid in its pores.

2. *Particle size exceeds 60 mm.* A rock or brick over this size can be debris. A clump of clay that is over this size would not be debris because the clay particles are smaller than 60 mm.

3. *Manufactured object* (e.g., bricks, glass, metal, concrete, paper, plastic, equipment), *or plant/animal matter* (e.g., tree stumps, branches, animal carcasses), *or natural geologic material* (e.g., rocks, boulders). Process wastes or residues cannot be debris.

§261.3(f)(1)   Hazardous debris that is treated by an appropriate extraction or destruction LDR treatment standard in 40 CFR 268.45 becomes non-hazardous under RCRA. Approval of the RCRA implementing agency is not required; however, documentation of proper treatment is necessary. Materials that do not meet the definition of debris are not eligible for this exclusion.

Spill Residues

Spill response and cleanup often produces contaminated residues and materials that need to be disposed of. Use TABLE 4-3 to help classify spill residues under RCRA.

Table 4-3. Spill Residue Classification

Material Spilled	Regulated As	Regulatory Citation
Waste listed solely for ignitability, corrosivity, or reactivity.	Hazardous only if it exhibits a hazardous characteristic.	§261.3 (g)(2)(ii)
Any other listed hazardous waste.	Listed hazardous waste.	§261.3 (c)(2)(i)
Material listed in §261.33(e)-(f).	Listed hazardous waste.	§261.33 (d)
Characteristic hazardous waste.	Hazardous only if it exhibits a hazardous characteristic.	§261.3 (d)(1)
Material *not* listed in §261.33(e), *or* which exhibits a characteristic.	Hazardous only if it exhibits a hazardous characteristic.	§261.3 (d)(1)(f)

CHAPTER 4: REVIEW QUESTIONS

Questions

1. A container is empty under RCRA if:

   ☐ A. there is no more than 1 inch of residue on the bottom
   ☐ B. no more material can be removed from the container using common practices
   ☐ C. visually one can see the bottom
   ☐ D. A and B

2. A container less than or equal to 119 gallons is RCRA empty if:

   ☐ A. residue is no more than 3% of the capacity of the container by *weight*
   ☐ B. residue is no more than 3% of the capacity of the container by *volume*
   ☐ C. there is no more than 1 inch of residue on the bottom
   ☐ D. A and C

3. A container that had been used to store K088 hazardous waste is emptied such that it is RCRA empty. Further residue is then removed from the container to make it more suitable for use. What is the RCRA status of the residue?

   ☐ A. K088
   ☐ B. not hazardous, unless it exhibits a hazardous characteristic
   ☐ C. not hazardous, regardless of if it exhibits a hazardous characteristic

4. Which of the following would be automatically regulated as hazardous wastes? (Check all that apply.)

   ☐ A. rinsings from container decontamination
   ☐ B. rinsings and liners from decontaminating a container of an acute hazardous chemical listed in 40 CFR 261.33(e)
   ☐ C. residue from a RCRA empty container
   ☐ D. a discarded RCRA empty container

5. The RCRA Empty Container Rule applies to which of the following?
   (Check all that apply.)

   ☐ A. 55-gallon drums
   ☐ B. roll-off containers
   ☐ C. tanks
   ☐ D. tanks on trucks
   ☐ E. 500 ml beakers

6. Listed hazardous waste is like King Midas: whatever it touches turns into hazardous waste.

   ☐ Truth or ☐ Myth?

7. A solid waste derived from treatment, storage, or disposal of a hazardous waste is a hazardous waste only if it exhibits a hazardous characteristic.

   ☐ Truth or ☐ Myth?

8. Which of the following may be subject to the Contained-In Policy or Rule?
   (Check all that apply.)

   ☐ A. a non-hazardous sludge that contains F001
   ☐ B. clothing contaminated with F006
   ☐ C. a dead frog that has F019 in its belly
   ☐ D. soil mixed with F019

9. Which of the following factors are relevant when determining whether a hazardous waste/wastewater mixture is subject to be regulated as hazardous wastes under RCRA?
   (Check all that apply.)

   ☐ A. weekly usage of solvents
   ☐ B. weekly generation of spent solvent
   ☐ C. wastewater's discharge status under the Clean Water Act
   ☐ D. wastewater flow rate
   ☐ E. the circumstances of how the hazardous waste enters the wastewater

Answers

1. ☒ A. A container is empty under RCRA if there is no more than 1 inch of residue on the bottom is only one of the two conditions that must be met for a container to be empty under RCRA.

   ☒ B. A container is empty under RCRA if no more material can be removed from the container using common practices is only one of the two conditions that must be met for a container to be empty under RCRA.

   ☒ C. A container is empty under RCRA if visually one can see the bottom is an incorrect statement.

   ☑ D. A container is empty under RCRA if the conditions in A and B are met: there is no more than 1 inch of residue on the bottom *and* all material has been removed using practices typically used for that type of container.

2. ☒ A. For a container less than or equal to 119 gallons, it is empty under RCRA if two conditions are met. Residue is no more than 3% of the capacity of the container by *weight* is only one condition.

   ☒ B. For a container less than or equal to 119 gallons, it is empty under RCRA if two conditions are met. Residue is no more than 3% of the capacity of the container by *volume* is not one of the conditions. The percent is correct, but the measurement is by weight not volume.

   ☒ C. For a container less than or equal to 119 gallons, it is empty under RCRA if two conditions are met. No more material can be removed from the container using common practices is only one condition.

   ☑ D. A container is empty under RCRA if the conditions in A and C are met: Residue is no more than 3% of the capacity of the container by *weight and* no more material can be removed from the container using common practices.

3. ☒ A. The RCRA status of the residue is K088 is an incorrect statement. A RCRA empty container is no longer regulated as hazardous waste.

   ☑ B. The RCRA status of the residue is not hazardous, unless it exhibits a hazardous characteristic.

   ☒ C. The RCRA status of the residue is not hazardous, regardless of if it exhibits a hazardous characteristic is an incorrect statement.

4. ☒ A. Rinsings from decontaminating a non-RCRA empty container would be automatically regulated as hazardous wastes is an incorrect statement. However, rinsings from container decontamination may be hazardous per the Derived-From Rule.

   ☑ B. Rinsings and liners from decontaminating a container of an acute hazardous chemical listed in 40 CFR 261.33(e) would be automatically regulated as hazardous wastes.

   ☒ C. Residue from a RCRA empty container would be automatically regulated as hazardous wastes is an incorrect statement. Residue from a RCRA empty container would be regulated as hazardous wastes only if it exhibits a hazardous characteristic.

   ☒ D. A discarded RCRA empty container would be automatically regulated as hazardous wastes is an incorrect statement. A discarded RCRA empty container would be regulated as hazardous wastes only if it exhibits a hazardous characteristic.

5. ☑ A. The RCRA Empty Container Rule applies to 55-gallon drums.

   ☑ B. The RCRA Empty Container Rule applies to roll-off containers.

   ☒ C. The RCRA Empty Container Rule applies to tanks is an incorrect statement. The Empty Container Rule applies to anything that meets the RCRA definition of "container" in 40 CFR 260.10:

   > "Container means any portable device in which a material is stored, transported, treated, disposed of, or otherwise handled."

   A key criterion is being portable, so stationary tanks are not considered containers. However, even a huge storage thing, such as a tote tank, is a container if it is moved around.

   ☑ D. The RCRA Empty Container Rule applies to tanks on trucks. Trucks move around and thus are portable.

   ☑ E. The RCRA Empty Container Rule applies to 500 ml beakers.

6. ☑ Myth

   Listed hazardous waste is like King Midas: whatever it touches turns into hazardous waste is a myth. Beliefs like this magnify the importance of understanding how the regulations actually work. If a listed waste "touches" something, the regulatory status of the "something" should be determined using the hazardous waste definition rules, including the Mixture and Derived-From Rules, and the Contained-In Policy and Rule.

7. ☑ Myth

    A solid waste derived from treatment, storage, or disposal of a hazardous waste is a hazardous waste only if it exhibits a hazardous characteristic is a myth. This is truth only if the hazardous waste is a characteristic hazardous waste or waste listed solely for ignitability, corrosivity, or reactivity.

8. ☒ A. A non-hazardous sludge that contains F001 may be subject to the Contained-In Policy or Rule is an incorrect statement. Non-hazardous sludge is neither environmental media nor debris, so the Contained-In Policy and Rule do not apply.

    ☑ B. Clothing contaminated with F006 is subject to the Contained-In Rule. The Contained-In Rule applies to "debris." Clothing meets the definition of "debris."

    ☑ C. A dead frog that has F019 in its belly is subject to the Contained-In Rule. The Contained-In Rule applies to "debris." The dead frog meets the definition of "debris."

    ☑ D. Soil mixed with F019 is subject to the Contained-In Policy. The Contained-In Policy applies to environmental media, which includes soil.

9. ☑ A. Weekly usage of solvents is a relevant factor when determining whether a hazardous waste/wastewater mixture is subject to be regulated as hazardous wastes under RCRA.

    ☒ B. Weekly generation of spent solvent is a relevant factor when determining whether a hazardous waste/wastewater mixture is subject to be regulated as hazardous wastes under RCRA is an incorrect statement.

    ☑ C. Wastewater's discharge status under the Clean Water Act is a relevant factor when determining whether a hazardous waste/wastewater mixture is subject to be regulated as hazardous wastes under RCRA.

    ☑ D. Wastewater flow rate is a relevant factor when determining whether a hazardous waste/wastewater mixture is subject to be regulated as hazardous wastes under RCRA.

    ☑ E. The circumstances of how the hazardous waste enters the wastewater is a relevant factor when determining whether a hazardous waste/wastewater mixture is subject to be regulated as hazardous wastes under RCRA.

CHAPTER 5:
Universal Wastes

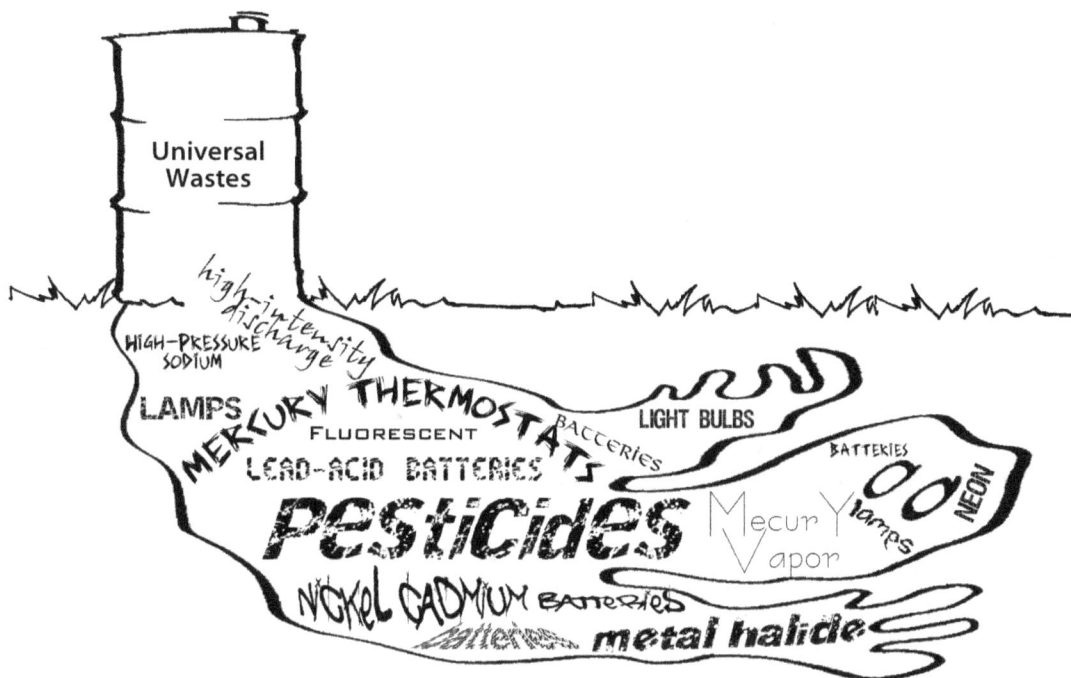

MANY COMMON ITEMS WIDELY used throughout society can become hazardous wastes when discarded. Four such items, called "universal wastes," are eligible for alternative management standards under §273: batteries, certain pesticides, mercury-containing equipment, and lamps.

To be a universal waste, a waste must first be a hazardous waste: many batteries, pesticides, mercury devices, and lamps are not hazardous wastes. The purpose of the universal wastes regulations is to encourage recycling. However, these regulations do not require recycling of universal wastes—they apply regardless of whether the universal wastes are shipped to a recycling facility or for disposal or treatment. *Also keep in mind that this is an optional set of regulations:* you may manage universal wastes under the full set of hazardous waste regulations if you choose.

Batteries

§273.2   Batteries are not specifically listed as hazardous waste, but they may exhibit one or more hazardous characteristics. Typically, nickel-cadmium and lead-acid batteries exhibit hazardous characteristics. Any discarded hazardous battery, regardless of size, type, or use, may be managed as a universal waste, provided it is "a device consisting of one or more electrically connected electrochemical cells which is designed to receive, store, and deliver electric energy."

§273.2(a)(2)   Spent lead-acid batteries may be managed either as universal waste or as lead-acid batteries under §266 Subpart G.

Pesticides

§273.3   Pesticides may either be listed in the P and U lists of 40 CFR 261.33(e)-(f) or exhibit the TC. The only type of pesticides that qualify as universal wastes are those that are subject to a recall or are being collected and managed during a waste pesticide collection program.

Mercury-Containing Equipment

§273.4   Mercury-containing equipment includes thermostats, switches, manometers, and other devices that contain metallic mercury, sometimes contained in a small vial. This equipment may exhibit the TC for mercury.

Lamps

§273.5   Any kind of electric light bulb may be considered a "lamp," including fluorescent, high-intensity discharge, neon, mercury vapor, high-pressure sodium, and metal halide. Lamps may exhibit the TC for mercury or other metals such as lead.

Small and Large Quantity Handlers

§273.9   Your handler size determines the universal waste requirements that apply to you, and it depends on the total amount of universal waste that you accumulate onsite at any one time. If you accumulate less than 5,000 kg of universal wastes at any time, you are a small quantity handler (SQH). As soon as you accumulate 5,000 kg or more, you are considered a large quantity handler (LQH). Your designation as LQH sticks to you through the calendar year, even if your accumulation falls below 5,000 kg.

GENERAL REQUIREMENTS

§273.11/§273.31
§273.12/§273.32

Both SQHs and LQHs are prohibited from disposing universal waste onsite and from treating universal waste in a way other than allowed under the specific waste management standards. LQHs must have an EPA ID number, but SQHs do not.

SPECIFIC WASTE MANAGEMENT STANDARDS

§273.13/§273.33
§273.14/§273.34

For each universal waste there are specific management standards for storage, preventing breakage and releases, performing certain onsite treatment activities, and labeling. These standards are the same for SQHs and LQHs.

ONSITE MANAGEMENT

§273.15/§273.35

Both SQHs and LQHs may accumulate universal wastes onsite for up to one year. For accumulation over a year, the SQH or LQH must be able to prove that the storage is necessary to accumulate enough to facilitate proper recycling, treatment, or disposal. The regulations require SQHs and LQHs to be able to demonstrate the length of time universal waste has been accumulated.

§273.16/§273.36

Both SQHs and LQHs must train employees about universal wastes, but the training requirements for LQHs are more stringent than those for SQHs. SQHs need only inform employees who handle universal waste about proper handling and emergency procedures. However LQHs "must ensure that all employees are thoroughly familiar with proper waste handling and emergency procedures, relative to their responsibilities during normal facility operations and emergencies."

§273.17/§273.37

Both SQHs and LQHs must respond to releases of universal wastes by "immediately" containing the wastes and residues. Also, materials generated from the cleanup must be characterized and managed under RCRA if hazardous.

OFFSITE MANAGEMENT

§273.18/§273.38
§273.19/§273.39

There are several requirements for transporting universal waste offsite that apply to both SQHs and LQHs. Hazardous waste manifests are not required for shipping universal wastes within the U.S. However, SQHs and LQHs must comply with applicable DOT hazardous material requirements. Anyone who transports universal wastes offsite must follow the Universal Waste Transporter Standards of 273 Subpart D.

LQHs are required to maintain records of universal waste shipments; SQHs are not. Exporting universal wastes from the U.S. must be done in compliance with the hazardous waste exporting requirements in 262, Subpart E.

CHAPTER 5: REVIEW QUESTIONS

Questions

1. The purpose of the universal waste regulations is to:
   (Check all that apply.)

   ☐ A. broaden the scope of wastes regulated as hazardous waste

   ☐ B. encourage recycling of commonly generated hazardous wastes

   ☐ C. encourage use of energy-saving batteries and light bulbs

   ☐ D. enable onsite disposal of certain wastes

2. Which of the following may be universal wastes?
   (Check all that apply.)

   ☐ A. batteries that contain heavy metals

   ☐ B. any out-of-date pesticide

   ☐ C. mercury-containing electrical equipment

   ☐ D. light bulbs that leach mercury and lead

3. LQHs of universal waste:
   (Check all that apply.)

   ☐ A. accumulate more than 5,000 kg of universal waste

   ☐ B. generate more than 5,000 kg of universal waste a month

   ☐ C. are exempt from having an EPA ID number

   ☐ D. are exempt from management standards if they don't generate hazardous waste

4. One can avoid becoming an LQH of universal wastes by:
   (Check all that apply.)

   ☐ A. making frequent enough shipments to ensure that less than 5,000 kg of universal wastes remains onsite at any one time

   ☐ B. managing universal wastes as hazardous wastes rather than as universal wastes

   ☐ C. making purchasing changes to ensure that no batteries, pesticides, mercury-containing equipment, or lamps used onsite exhibit a hazardous waste characteristic when discarded

☐ D. applying for an exemption for LQH status

5. Shipments of universal wastes:
    (Check all that apply.)

☐ A. require a hazardous waste manifest

☐ B. are exempt from DOT regulations

☐ C. are subject to recordkeeping requirements

☐ D. may not leave the U.S.

Answers

1. ☒ A. The purpose of the universal waste regulations is to broaden the scope of wastes regulated as hazardous waste is an incorrect statement. Wastes subject to regulation as universal wastes already meet the definition of hazardous waste.

   ☑ B. One purpose of the universal waste regulations is to encourage recycling of commonly generated hazardous wastes. The universal waste regulations provide an alternative and more flexible set of regulations to manage common items that are hazardous when disposed of.

   ☒ C. The purpose of the universal waste regulations is to encourage use of energy-saving batteries and light bulbs is an incorrect statement.

   ☒ D. The purpose of the universal waste regulations is to enable onsite disposal of certain wastes is an incorrect statement. Disposal onsite of universal wastes is prohibited unless you have a hazardous waste disposal permit.

2. ☑ A. Batteries that contain heavy metals may be universal wastes.

   ☒ B. Any out-of-date pesticide may be universal wastes is an incorrect statement because the only pesticides eligible for management as universal waste are those that are being recalled or are part of a collection program for pesticides.

   ☑ C. Mercury-containing electrical equipment may be universal wastes.

   ☑ D. Light bulbs that leach mercury and lead may be universal wastes.

3. ☑ A. LQHs of universal waste accumulate more than 5,000 kg of universal waste.

   ☒ B. LQHs of universal waste generate more than 5,000 kg of universal waste a month is an incorrect statement. Handler status depends on the quantity of universal waste *accumulated* not generated.

   ☒ C. LQHs of universal waste are exempt from having an EPA ID number is an incorrect statement. An LQH of universal waste is required to have an EPA ID number.

   ☒ D. LQHs of universal waste are exempt from management standards if they don't generate hazardous waste is an incorrect statement. LQHs of universal waste are subject to universal waste management standards regardless of their status as a hazardous waste generator.

4. ☑ A. One can avoid becoming an LQH of universal wastes by making frequent enough shipments to ensure that less than 5,000 kg of universal wastes remains onsite at any one time. Since handler status depends on amount *accumulated* and not generated, scheduling shipments of universal waste to maintain less than 5,000 kg onsite is a common strategy to maintain SQH status.

☑ B. One can avoid becoming an LQH of universal wastes by managing universal wastes as hazardous wastes rather than as universal wastes. One certainly has the option of managing universal wastes as hazardous wastes. The universal waste regulations are an alternative to managing these materials as hazardous waste.

☑ C. One can avoid becoming an LQH of universal wastes by making purchasing changes to ensure that no batteries, pesticides, mercury-containing equipment, or lamps used onsite exhibit a hazardous waste characteristic when discarded. The universal waste regulations apply only to hazardous wastes, so if your spent batteries, pesticides, or mercury-containing equipment and lamps are not hazardous wastes, they are also not universal wastes.

☒ D. One can avoid becoming an LQH of universal wastes by applying for an exemption for LQH status is an incorrect statement. There is no exemption petition process for handler status.

5. ☒ A. Shipments of universal wastes require a hazardous waste manifest is an incorrect statement. Universal waste shipments are exempt from manifesting.

☒ B. Shipments of universal wastes are exempt from DOT regulations is an incorrect statement. DOT regulations do apply to universal waste shipments.

☑ C. Shipments of universal wastes are subject to recordkeeping requirements. LQHs must maintain records of shipments.

☒ D. Shipments of universal wastes may not leave the U.S. is an incorrect statement. Universal wastes may be exported; however, the hazardous waste exporting regulations apply.

CHAPTER 6:
Hazardous Wastes
Used, Reused, Reclaimed, or Recycled

I S IT A WASTE if you can use it, reuse it, reclaim it, or recycle it? That depends on the nature of the waste and how the materials are handled. This chapter provides a six-step process that will help you determine how your recycled waste is regulated under RCRA. After working through this process, if your material is regulated as "waste," you must manage that material as hazardous waste.

Six Steps to Determine How a Recycled Hazardous Waste Is Regulated Under RCRA

§261.2(c)-(d)

Under RCRA, recycled hazardous wastes are regulated as hazardous wastes depending on their nature and how the materials are handled. To determine how a recycled hazardous waste is regulated under RCRA, follow these six steps:

1. Classify as secondary material

2. Determine if material is "inherently waste-like"

3. Classify recycling activity

4. Determine status

5. Document exemption claims

6. Evaluate variance and non-waste determination eligibility

If after working through this process, your material is regulated as "waste," you must manage that material as hazardous waste.

STEP 1: CLASSIFY AS SECONDARY MATERIAL

§261.1(c)

RCRA defines the following types of secondary materials: spent material, sludge, by-product, commercial chemical product, hazardous scrap metal, and excluded scrap metal.

Spent Material

Spent material is material that from use has become contaminated by physical or chemical impurities and can no longer serve the purpose for which it was produced without processing or reclaiming it. It may not be "spent" if it is used in another process effectively without first being reclaimed. For example, spent solvents and spent acids and caustics.

Sludge

Sludge is material generated from air or water treatment or pollution control equipment, exclusive of the treated effluent. For example, chromium hydroxide sludge from metal finishing wastewater treatment and spent carbon from decontaminating groundwater.

By-product

By-product is not a primary product and not solely or separately produced by the production process—by-product is not useful without processing. For example, metal drosses and slags.

Commercial Chemical Product

Commercial chemical product or intermediate that is either listed in 40 CFR 261.33 or not listed but exhibits a hazardous characteristic, which is off-spec, spill residue, or otherwise unwanted. For example, pesticide that contains endrin as active ingredient discarded because it is out of date.

Hazardous Scrap Metal

Hazardous scrap metal is bits and pieces of metal parts or metal pieces that can be recycled. For example, shavings of metallic lead that exhibit the TC, radiators, scrap automobiles, and aerosol cans.

Excluded Scrap Metal

Excluded scrap metal is metal that has been processed (for example, baled, shredded, sheared, chopped, crushed, flattened, cut, melted, and sorted), agglomerated metal fines or drosses, and metal turnings, cuttings, punchings, and borings.

Excluded scrap metal being recycled is not solid waste.

STEP 2: DETERMINE IF MATERIAL IS INHERENTLY WASTE-LIKE

§261.2(d)

Inherently waste-like materials are listed by EPA through rulemaking using specific criteria. This list, §261.2(d), now contains: F020, F021, F022, F023, F026, F028, and hazardous waste fed to a halogen acid furnace.

All inherently waste-like materials are regulated as wastes.

STEP 3: CLASSIFY RECYCLING ACTIVITY

Classify the recycling activity as follows: use constituting disposal, burning for energy recovery or use as fuel, reclamation, speculative accumulation, use as a feedstock or ingredient, closed loop recycling, reclamation under the control of the generator, or transfer to another person for reclamation.

Use Constituting Disposal

§261.2(c)(1)

Applying to or placing on the land *or* contained in products that are applied to the land.

The intent of use constituting disposal is to regulate those wastes that are land disposed under the guise of soil or land beneficiation. All secondary materials are solid wastes if applied to the land, except commercial chemical products whose ordinary use is land application, for example, pesticides.

Burning for Energy Recovery or Use as Fuel

§261.2(c)(2)

All secondary materials are solid wastes if burned to recover energy or used as fuel, except commercial chemical products whose ordinary use is as a fuel.

Reclamation

§261.1(c)(4)

"Reclamation" is defined as a process that either recovers a usable product from waste or regenerates waste to make it usable. Burning *solely* for material recovery is considered reclamation. For example, redistilling spent solvents, regenerating carbon filters, and recovering cryolite from spent potlining.

Speculative Accumulation

§261.1(c)(8)

Speculative accumulation is storing waste in anticipation of eventually recycling or using it, but without recycling or using a significant portion of the stored waste. If 75% of a waste in storage *on January 1* is not recycled *within the calendar year*, the storage is considered "speculative accumulation."

All secondary materials, except commercial chemical products, are regulated as solid wastes if their storage is considered to be speculative accumulation.

Use as a Feedstock or Ingredient

§261.2(e)

A secondary material is used as a feedstock or ingredient *if it is:*

> ➤ used or reused as an ingredient in an industrial process to make a product, *or*

> ➤ used or reused as a substitute for a commercial product, *or*

> ➤ returned to the original process as a substitute for a raw material;

and it is not:

> ➤ reclaimed before use, *or*

➤ stored as speculative accumulation, *or*

➤ incorporated into a fuel, or burned as fuel, *or*

➤ incorporated into a product that is applied to the land, *or*

➤ inherently waste-like.

A secondary material used as a feedstock or ingredient is not regulated as a waste.

The scope of this category is narrow, so be sure you understand your situation and the above provisions before claiming this exemption.

Closed Loop Recycling

§261.4(a)(8) A secondary material is recycled by closed loop recycling *if it is:*

➤ reclaimed and returned to the same production process that generated it, *and*

➤ managed entirely in tanks and piping or other enclosed way;

and it is not:

➤ reclaimed in a way involving combustion of the waste, *or*

➤ accumulated longer than a year without being reclaimed, *or*

➤ incorporated into a fuel, or burned as fuel, *or*

➤ incorporated into a product that is applied to the land.

A secondary material managed in closed loop recycling is not regulated as a waste.

Reclamation Under the Control of the Generator

§261.2(a)(ii) A secondary material is generated and reclaimed under the control of the generator *if it is:*

➤ handled only in *non-land-based units* and is contained in such units, *and*

➤ generated and reclaimed within the U.S. and its territories, *and*

➤ in conformance with the §260.43 criteria, *and*

➤ notified under §260.42;

and it is not:

➤ speculatively accumulated, *or*

➤ a spent lead acid battery, *or*

➤ K171 or K172, *or*

➤ subject to material-specific management conditions under §261.4(a) when reclaimed.

The exemption at 40 CFR 261.2(a)(ii) applies only to wastes managed in "non-land-based units." Examples of non-land-based units are containers, tanks, containment buildings, and impoundments used for mineral processing.

§261.4(a)(23)   *In addition to the above requirements*, hazardous secondary materials managed in *land-based units*, must be "contained." The exemption for hazardous secondary materials managed in "land-based units" at 261.4(a)(23) defines "land-based unit" as:

> "An area where hazardous secondary materials are placed in or on the land before recycling. This definition does not include land-based production units."

Examples of land-based units are surface impoundments, landfills, and waste piles.

A secondary material generated and reclaimed under the control of the generator is not regulated as waste.

Transfer to Another Person for Reclamation

§261.4(a)(24)   A secondary material is transferred to another person for reclamation *if it is:*

➤ handled only by the generator, transporter, intermediate facility, and reclaimer, *and*

➤ stored less than 10 days at a transfer facility, *and*

➤ packaged in compliance with DOT regulations during transportation, *and*

➤ in conformance with the §260.43 criteria, *and*

➤ notified under §260.42, *and*

➤ contained;

and it is not:

➤ speculatively accumulated, *or*

➤ a spent lead acid battery, *or*

➤ K171 or K172, *or*

➤ subject to material-specific management conditions under §261.4(a) when reclaimed.

The generator must:

➤ file notification under §260.42, *and*

➤ perform a due diligence of each intermediate facility and reclaimer that will handle the secondary material for specific conditions [§261.4(a)(24)(v)(B)], *and*

➤ maintain documentation and certification [§261.4(a)(24)(v)(C)], *and*

➤ maintain records of shipments and confirmations of receipt [§261.4(a)(24)(v)(D)(E)].

The intermediate facility must:

➤ send the hazardous secondary material to the reclaimer(s) designated by the generator [§261.4(a)(24)(vi)(B)].

The intermediate facility and reclaimer must:

➤ file notification under §260.42, *and*

➤ maintain records of shipments received [§261.4(a)(24)(vi)(A)], *and*

➤ send to generator confirmations of receipt for each shipments received [§261.4(a)(24)(vi)(C)], *and*

➤ contain secondary material and manage it "at least as protective as that employed for analogous raw material" [§261.4(a)(24)(vi)(D)], *and*

➤ manage residuals in a "manner that is protective of human health and the environment" [§261.4(a)(24)(vi)(E)], *and*

➤ manage residual hazardous wastes in compliance with hazardous waste regulations [§261.4(a)(24)(vi)(E)], *and*

➤ maintain financial assurance [§261.4(a)(24)(vi)(F)].

A secondary material transferred to another person for reclamation is not regulated as waste.

STEP 4: DETERMINE STATUS

If you know the type of material and how the material is being recycled, you can use the *RCRA Hazardous Waste Matrix* in FIGURE 6-1, page 78, to determine if the material is regulated as a RCRA hazardous waste.

Type of Material	How the Material Is Being Recycled*			
	Use Constituting Disposal	Use as Fuel	Reclamation	Speculative Accumulation
Spent Material	Waste	Waste	Waste	Waste
Listed Sludge	Waste	Waste	Waste	Waste
Characteristic Sludge	Waste	Waste	Not Waste	Waste
Listed By-product	Waste	Waste	Waste	Waste
Characteristic By-product	Waste	Waste	Not Waste	Waste
Commercial Chemical Product	Waste	Waste	Not Waste	Not Waste
Scrap Metal Not Excluded by §261.1(c)(9)	Waste	Waste	Exempt	Waste

Source: §261.2, Table 1.

*Waste: Defined as waste.
Not Waste: Not defined as waste.
Exempt: Defined as waste but exempt [§261.6(a)(3)(ii)].

Figure 6-1. RCRA Hazardous Waste Matrix

STEP 5: DOCUMENT EXEMPTION CLAIMS

If you determine that a waste is not defined as RCRA solid waste, you must carefully document the claim. Documentation should include a detailed description of how the waste is recycled and why it is exempted. The section *Waste vs. Raw Material: Demonstrations Criteria*, page 79, has guidelines for evaluating claims and what to include in the documentation.

STEP 6: EVALUATE VARIANCE AND NON-WASTE DETERMINATION ELIGIBILITY

If you determine that your waste is regulated as waste, you should evaluate the specific situations where an EPA Regional Administrator or authorized state may grant a variance or non-waste determination.

Variances

§260.31(a) ◼ Secondary material is accumulated speculatively, but is expected to be recycled within the next year.

§260.31(b) ◼ Secondary material is reclaimed, then used as feedstock within the original production process. The reclamation process must be an essential part of the production process.

§260.31(c) ◼ Secondary material has been reclaimed but must be reclaimed further before being used. The materials must be "commodity-like" after the first reclamation.

Non-waste Determinations

§260.34(b) ◼ Secondary material that is reclaimed in a continuous industrial process.

§260.34(c) ◼ Secondary material that is indistinguishable in all relevant aspects from a product or an intermediate.

If, after working through this process, your material is regulated as "waste," you must manage that material as hazardous waste.

Waste vs. Raw Material: Demonstrations Criteria

For years, non-promulgated guidelines were relied upon to make crucial decisions of whether a material was a waste or a raw material. Now there are promulgated criteria for these decisions. The §260.43 Criteria are promulgated criteria required for demonstrating that a specific recycling case is legitimate recycling and not sham recycling. Also listed here are the criteria used for non-waste determinations.

DEMONSTRATING LEGITIMATE RECYCLING (§260.43 CRITERIA)

Legitimacy Criteria That Must Be *Addressed*

1. Hazardous secondary material provides a useful contribution *if it:*

   ➤ contributes valuable ingredients to a product or intermediate, *or*

   ➤ replaces a catalyst or carrier in the recycling process, *or*

   ➤ is the source of a valuable constituent recovered in the recycling process, *or*

   ➤ is recovered or regenerated by the recycling process, *or*

   ➤ is used as an effective substitute for a commercial product.

2. Recycling process produces a valuable product or intermediate *if it is:*

> ➤ sold to a third party, *or*

> ➤ used by the recycler or the generator as an effective substitute for a commercial product or as an ingredient or intermediate in an industrial process.

Legitimacy Criteria That Must Be *Considered*

1. Hazardous secondary material is managed as a valuable commodity *such that:*

> ➤ where there is an analogous raw material, the hazardous secondary material is managed, at a minimum, in a manner consistent with the management of the raw material, *and*

> ➤ where there is no analogous raw material, the hazardous secondary material is contained, *and*

> ➤ hazardous secondary materials that are released to the environment and are not immediately recovered are discarded.

2. The product of the recycling process *does not:*

> ➤ contain significant concentrations of any Appendix VIII hazardous constituents that are not in analogous products, *or*

> ➤ contain any Appendix VIII hazardous constituents at levels that are significantly elevated from those found in analogous products, *or*

> ➤ exhibit a hazardous characteristic that analogous products do not exhibit.

In evaluating a hazardous secondary material, *wholeness counts* as both factors must be evaluated and *legitimacy as a whole considered*. If one or both factors are *not* met, this may indicate that the material is not legitimately recycled.

These two factors must be *considered*, but they do not necessarily have to be *met* for the recycling to be considered legitimate. *If one or both of these factors are not met, the following criteria can be considered:*

> ➤ protectiveness of the storage methods, *and*

> ➤ exposure from toxics in the product, *and*

> ➤ the bioavailability of the toxics in the product, *and*

> ➤ other relevant considerations.

NON-WASTE DETERMINATIONS (§260.34 CRITERIA)

Secondary Material Is Reclaimed in a Continuous Industrial Process [§260.34(b)]

§260.43 criteria, and

➤ hazardous secondary material is part of the continuous primary production process and not waste treatment, *and*

➤ capacity of the production process will use the hazardous secondary material in a reasonable time frame, *and*

➤ indicators that the hazardous secondary material will not be abandoned (for example, past practices, market factors, the nature of the hazardous secondary material, and contractual arrangements), *and*

➤ hazardous constituents in the hazardous secondary material are reclaimed rather than released to the air, water, or land at significantly higher levels (either statistically or risk-wise) than would otherwise be released by the production process, *and*

➤ other relevant factors that demonstrate the hazardous secondary material is not discarded.

Secondary Material Is Indistinguishable in All Relevant Aspects from a Product or Intermediate [§260.34(c)]

§260.43 criteria, and

➤ indicators that market participants treat the hazardous secondary material as a raw material rather than a waste (for example, current positive value of the hazardous secondary material, stability of demand, and contractual arrangements), *and*

➤ the hazardous secondary material is chemically and physically comparable to commercial products or intermediates, *and*

➤ capacity of the market will use the hazardous secondary material in a reasonable time frame, *and*

➤ indicators that the hazardous secondary material will not be abandoned (for example, past practices, market factors, the nature of the hazardous secondary material, and contractual arrangements), *and*

➤ hazardous constituents in the hazardous secondary material are reclaimed rather than released to the air, water, or land at significantly higher levels (either statistically or risk-wise) than would otherwise be released by the production process, *and*

➤ other relevant factors that demonstrate the hazardous secondary material is not discarded.

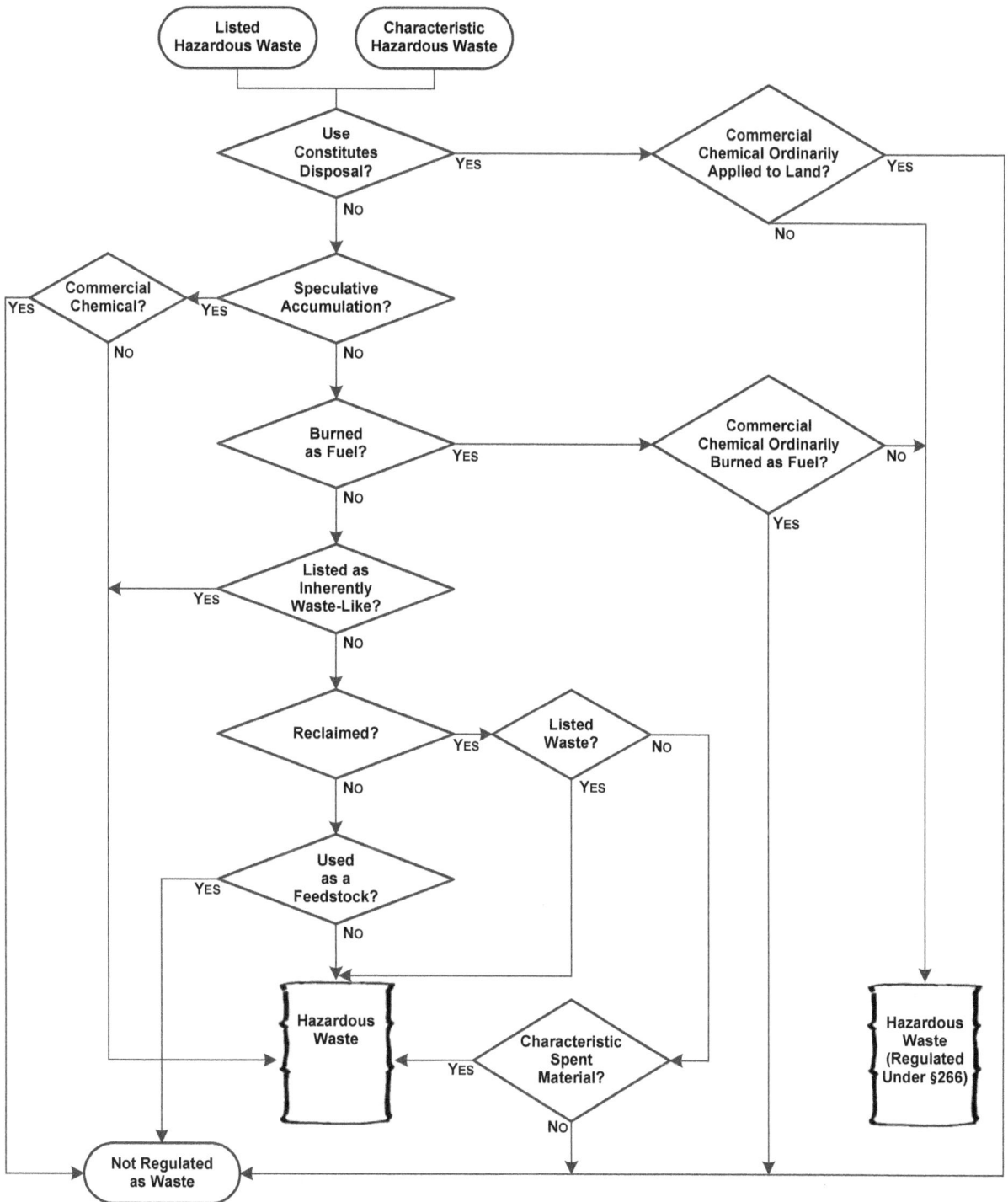

Figure 6-2. Simplified Look at Hazardous Waste Used, Reused, Reclaimed, or Recycled

CHAPTER 6: REVIEW QUESTIONS

Questions

1. If I recycle a hazardous waste, it ceases to be a waste and isn't regulated as hazardous waste under RCRA.

    ☐ Truth or ☐ Myth?

2. Which of the following terms describe "secondary materials" under RCRA?
    (Check all that apply.)

    ☐ A. spent material
    ☐ B. sludge
    ☐ C. by-product
    ☐ D. debris

3. Determinations of whether a waste is "inherently waste-like" are made on a case-by-case basis.

    ☐ Truth or ☐ Myth?

4. Which of the following describes hazardous waste recycling activities that may be exempt under RCRA?
    (Check all that apply.)

    ☐ A. closed loop recycling
    ☐ B. use as feedstock
    ☐ C. reclamation under control of generator
    ☐ D. reselling as consumer product

5. Which of the following "secondary materials" *may* be exempt as solid waste even if accumulated speculatively?
    (Check all that apply.)

    ☐ A. commercial chemical product
    ☐ B. spent material
    ☐ C. by-product
    ☐ D. sludge

6. Hazardous secondary materials managed in a land-based unit, such as a surface impoundment, are not eligible for exemption as solid waste due to the risk of releases from the unit.

☐ Truth or ☐ Myth?

7. Which of the following is true regarding hazardous waste transferred to another person for reclamation?
(Check all that apply.)

☐ A. DOT regulations must be followed

☐ B. notifications are required

☐ C. legitimacy criteria of 40 CFR 260.43 apply

☐ D. reclamation must be approved by a RCRA-authorized government agency

8. Which of the following is true when evaluating whether a hazardous waste exists under the legitimacy criteria of 40 CFR 260.43?
(Check all that apply.)

☐ A. all criteria must be addressed and satisfied

☐ B. some criteria must be addressed, and some criteria must only be considered

☐ C. the criteria are used in non-waste determinations

☐ D. documentation of criteria compliance is required

☐ E. documentation of criteria compliance must be approved by a RCRA-authorized government agency

9. Which of the following situations are eligible for a variance or non-waste determination from a RCRA-authorized government agency?
(Check all that apply.)

☐ A. waste that is being accumulated speculatively but provisions have been made for recycling

☐ B. waste that can be used as a feedstock in the production process that it came from if the waste is first reclaimed

☐ C. waste that is landfilled in a monofill that may be useful in the future

☐ D. waste that can substitute for a commercial product in an industrial process

10. Based on what you've read, which statement is the most true about the RCRA definition of solid waste?

☐ A. hazardous waste is not a waste if I "declare" that it is not

☐ B. once a material is considered "waste" it's always waste

☐ C. there are some exemptions available from the definition of solid waste, but the matter is not trivial

☐ D. if I can show that my waste doesn't pose a risk when recycled, it's not a RCRA solid waste

Answers

1. ☑ Myth

   If I recycle a hazardous waste, it ceases to be a waste and isn't regulated as hazardous waste under RCRA is a myth. Determination of a recycled hazardous waste's status under RCRA is one of the most complex aspects of waste characterization, depending on the waste itself and how it is recycled.

2. ☑ A. Spent material describes "secondary materials" under RCRA.

   ☑ B. Sludge describes "secondary materials" under RCRA.

   ☑ C. By-product describes "secondary materials" under RCRA.

   ☒ D. Debris describes "secondary materials" under RCRA is an incorrect statement.

3. ☑ Myth

   Determinations of whether a waste is "inherently waste-like" are made on a case-by-case basis is a myth—they *are not* made on a case-by-case basis. The only wastes considered to be "inherently waste-like" are listed in 40 CFR 261.2(d).

4. ☑ A. Closed loop recycling describes a hazardous waste recycling activity that may be exempt under RCRA.

   ☑ B. Use as feedstock describes a hazardous waste recycling activity that may be exempt under RCRA.

   ☑ C. Reclamation under control of generator describes a hazardous waste recycling activity that may be exempt under RCRA.

   ☒ D. Reselling as consumer product describes a hazardous waste recycling activity that may be exempt under RCRA is an incorrect statement.

5. ☑ A. Commercial chemical product is a secondary material that may be exempt as solid waste even if accumulated speculatively. For example, a drum of unused trichloroethylene that has become unusable in your process because it sat around too long may be speculatively accumulated as you try to find a home for it.

   ☒ B. Spent material is a secondary material that may be exempt as solid waste even if accumulated speculatively is an incorrect statement.

   ☒ C. By-product is a secondary material that may be exempt as solid waste even if accumulated speculatively is an incorrect statement.

☒ D. Sludge is a secondary material that may be exempt as solid waste even if accumulated speculatively is an incorrect statement.

6. ☑ Myth

Hazardous secondary materials managed in a land-based unit, such as a surface impoundment, are not eligible for exemption as solid waste due to the risk of releases from the unit is a myth. Hazardous secondary materials managed in a land-based unit may be exempt under 40 CFR 261.4(a)(23), which includes provisions requiring materials to be contained within the unit.

7. ☑ A. When transferring hazardous waste to another person for reclamation, DOT regulations must be followed.

   ☑ B. When transferring hazardous waste to another person for reclamation, notifications are required.

   ☑ C. When transferring hazardous waste to another person for reclamation, the legitimacy criteria of 40 CFR 260.43 apply.

   ☒ D. When transferring hazardous waste to another person for reclamation, the reclamation must be approved by a RCRA-authorized government agency is an incorrect statement. The activity does not require approval.

8. ☒ A. When evaluating whether a hazardous waste exists under the legitimacy criteria of 40 CFR 260.43, all criteria must be addressed and satisfied is an incorrect statement. Some criteria must be addressed, and some criteria must only be considered.

   ☑ B. When evaluating whether a hazardous waste exists under the legitimacy criteria of 40 CFR 260.43, some criteria must be addressed, and some criteria must only be considered.

   ☑ C. When evaluating whether a hazardous waste exists under the legitimacy criteria of 40 CFR 260.43, the criteria are used in non-waste determinations.

   ☑ D. When evaluating whether a hazardous waste exists under the legitimacy criteria of 40 CFR 260.43, documentation of criteria compliance is required.

   ☒ E. When evaluating whether a hazardous waste exists under the legitimacy criteria of 40 CFR 260.43, documentation of criteria compliance must be approved by a RCRA-authorized government agency is an incorrect statement. The documentation does not require approval.

9. ☑ A. Waste that is being accumulated speculatively but provisions have been made for recycling is a situation that is eligible for a variance or non-waste determination from a RCRA-authorized government agency.

☑ B. Waste that can be used as a feedstock in the production process that it came from if the waste is first reclaimed is a situation that is eligible for a variance or non-waste determination from a RCRA-authorized government agency.

☒ C. Waste that is landfilled in a monofill that may be useful in the future is a situation that is eligible for a variance or non-waste determination from a RCRA-authorized government agency is an incorrect statement. There is no variance from solid waste definition for wastes being land disposed.

☑ D. Waste that can substitute for a commercial product in an industrial process is a situation that is eligible for a variance or non-waste determination from a RCRA-authorized government agency.

10. ☒ A. Hazardous waste is not a waste if I "declare" that it is not is the most true statement about the RCRA definition of solid waste is incorrect. There's much more to the definition of solid waste than a generator declaration.

☒ B. Once a material is considered "waste" it's always waste is an incorrect statement. The regulations *do provide* mechanisms for exempting hazardous wastes that are used, reused, reclaimed, or recycled under certain conditions.

☑ C. There are some exemptions available from the definition of solid waste, but the matter is not trivial is the most true statement about the RCRA definition of solid waste.

☒ D. If I can show that my waste doesn't pose a risk when recycled, it's not a RCRA solid waste is the most true statement about the RCRA definition of solid waste is incorrect. Although there is some risk analysis involved, there are many other factors.

CHAPTER 7:
Delisting

ONCE A LISTED WASTE, always a listed waste—unless delisted. The delisting process is the waiver mechanism, delegated to EPA Regions and authorized states, to exclude specific wastes, by generator and site, from regulation as hazardous waste. The delisting process may also be used to delist waste contained in a site such as a landfill, surface impoundment, or spill residue.

Regulatory agencies take delisting very seriously since granting a delisting removes a waste from hazardous waste control by amending RCRA regulations. Federal delistings are first published as proposed rule in the *Federal Register* (FR) for public notice and comment. The EPA Region then issues a final rulemaking decision in the FR that adds the waste to 40 CFR 261 Appendix IX by specific waste, generator, and address.

RCRA's listing mechanism for defining hazardous wastes is a broad and fine mesh net that catches many wastes and automatically labels them as hazardous waste. Because many of the listing descriptions are designed to be broad, you may generate a waste that matches a listing description, but because of your raw materials, manufacturing process, waste treatment process, or other reasons, your particular waste may not be hazardous. This chapter discusses the delisting petition and delisting criteria, and it explains the EPA Delisting Model, so you can assess the delisting potential for your specific wastes.

Delisting Petition

§260.22

A delisting petition contains the following: (1) administrative information, (2) waste description and waste management history, (3) detailed description of the waste generating process and raw materials, and if applicable, the waste treatment system, (4) description of the constituents and properties of concern, (5) extensive sampling and analysis data, and (6) groundwater monitoring data (if applicable).

Delisting Criteria

Generally, a waste is eligible for delisting if it meets four criteria:

1. **The Waste Does Not Exhibit Any Hazardous Characteristics**

   No matter what the waste was listed for, if it exhibits any of the four hazardous characteristics, it is not eligible for delisting. For example, if your electroplating sludge (F006, listed for heavy metals and cyanide) exhibits reactivity, regardless of its heavy metal and cyanide content, it is not delistable.

2. **Acceptable Risk Results from EPA's Delisting Risk Assessment Software**

   EPA's Delisting Risk Assessment Software (DRAS) uses a model called EPA Composite Model for Leachate Migration with Transformation Products (EPACMTP) to predict the potential for a waste to affect the environment if delisted. This model uses the waste's annual generation rate and leachate and composition data of all hazardous constituents that may be in the waste; assumes disposal of the waste in a non-hazardous waste landfill or surface impoundment and worst case migration, fate, and transport of waste constituents; and calculates contaminant concentrations and cumulative risk at hypothetical receptor points. If the concentrations of contaminants are below established health-based standards, such as drinking water standards, the waste may be delistable.

3. **The Waste Has Not Caused Groundwater Contamination or Other Problems**

   If the waste is managed in an onsite or dedicated offsite land disposal facility, and if groundwater monitoring data shows contamination from the waste, then delisting is unlikely.

4. **The Waste Does Not Pose Other Hazards**

   When evaluating delistings, EPA Regions and authorized states have authority under RCRA to consider any other factors, such as presence of waste constituents that are not regulated as hazardous but do pose environmental or health threats.

Delisting Model

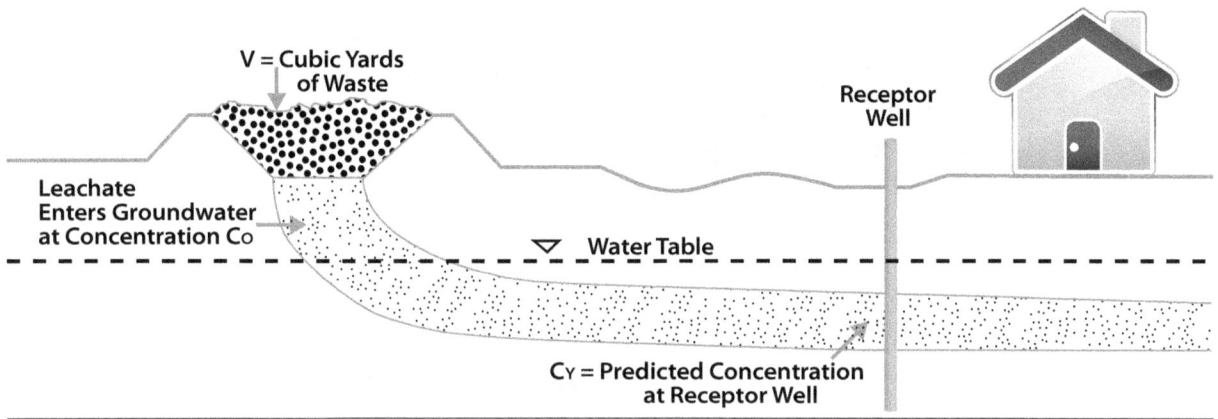

Figure 7-1. Simplified Look at EPA's Delisting Model

EPACMTP was developed by EPA to predict migration of hazardous constituents from a land disposal facility to a receptor drinking water well. EPA used an older version of the groundwater pathway for this model to set regulatory levels for the TC (see Chapter 2: *Hazardous Waste Definition: Four Hazardous Waste Characteristics: 4. Toxicity Characteristic*, page 21). This model was modified for evaluating wastes for delisting. Using input data for aquifer characteristics and other parameters based on surveys, field measurements, and literature sources, EPA modeled releases from two land disposal facility types: a non-hazardous waste landfill for physical solids and a surface impoundment for liquids. The model predicted receptor well concentrations based on a waste's leachate results and the volume of waste disposed as follows:

C_O: Concentration of contaminant in waste leachate

C_Y: Predicted concentration of contaminant at receptor

V: Waste annual generation volume

C_O/C_Y: Dilution/attenuation factor (DAF)

A DAF is an estimate of the reduction in concentration of a contaminant that leaches from a disposal facility and travels to a down-gradient drinking water well. The model considers reductions due to vertical and horizontal dispersion, sorption, chemical degradation, and other factors.

The type of disposal method—landfill or surface impoundment—is chosen based on current disposal methods and physical state of the waste.

Waste Delisting Potential Evaluation

Delisting takes time and money. Before launching into a delisting effort, do some homework. First, evaluate your waste for delisting potential by reviewing the four delisting criteria discussed on pages 92–93. Serious evaluation of a waste requires the use of DRAS[1] and a copy of the delisting guidance manual.[2] Contact your EPA Regional Office and state agency about delisting requirements for your specific waste. A little time up front can save you time and money later.

1. DRAS v3.0 download (http://www.epa.gov/region05/waste/hazardous/delisting/dras-software.html). *User's Guide* DRAS v3.0 (http://www.epa.gov/region05/waste/hazardous/delisting/pdfs/dras-uguide-200810.pdf); October 2008.

2. *EPA RCRA Delisting Program Guidance Manual for the Petitioner* (http://www.epa.gov/region6/6pd/rcra_c/pd-o/delist23.pdf); March 23, 2000.

CHAPTER 7: REVIEW QUESTIONS

Questions

1. A waste may be delisted if it:
   (Check all that apply.)

   ☐ A. doesn't exhibit any characteristics
   ☐ B. passes EPA groundwater model
   ☐ C. hasn't contaminated groundwater
   ☐ D. does not contain hazardous constituents in levels of concern
   ☐ E. all of the above

2. A hazardous waste that is delisted is not subject to any requirements under RCRA.

   ☐ Truth or ☐ Myth?

3. Which of the following statements are correct about delistings?
   (Check all that apply.)

   ☐ A. delisting is a self-implementing process
   ☐ B. delisting becomes effective after a regulatory official reviews the petition and issues an "intent to delist"
   ☐ C. delisting is time-consuming
   ☐ D. delisting is a good option for characteristic wastes that have been rendered non-hazardous through treatment

Answers

1. ☒ A. A waste may be delisted if it doesn't exhibit any characteristics is an incorrect statement because it is incomplete.

    ☒ B. A waste may be delisted if it passes EPA groundwater model is an incorrect statement because it is incomplete.

    ☒ C. A waste may be delisted if it hasn't contaminated groundwater is an incorrect statement because it is incomplete.

    ☒ D. A waste may be delisted if it does not contain hazardous constituents in levels of concern is an incorrect statement because it is incomplete.

    ☑ E. A waste may be delisted if it doesn't exhibit any characteristics, passes EPA groundwater model, hasn't contaminated groundwater, and does not contain hazardous constituents in levels of concern. The waste needs to meet all of these criteria and more.

2. ☑ Myth

    A hazardous waste that is delisted is not subject to any requirements under RCRA is a myth. A hazardous waste that is delisted is subject to the requirements established in 40 CFR 261 Appendix IX as conditions for its delisting.

3. ☒ A. Delisting is a self-implementing process is an incorrect statement. Delisting is implemented by a RCRA-authorized government agency.

    ☒ B. Delisting becomes effective after a regulatory official reviews the petition and issues an "intent to delist" is an incorrect statement. Delisting becomes effective when promulgated as rulemaking by a RCRA-authorized government agency.

    ☑ C. Delisting is time-consuming. The delisting process is now well-defined; however, the steps—including data collection, sampling/analysis, and petition preparation and review—can take months or even years.

    ☒ D. Delisting is a good option for characteristic wastes that have been rendered non-hazardous through treatment is an incorrect statement. Delisting doesn't apply to this waste. Delisting applies to listed hazardous wastes.

    Characteristic hazardous wastes that are no longer hazardous are no longer hazardous wastes. However, the treatment of the waste may be subject to RCRA regulation and there may be some requirements that attach to the waste because it was once hazardous, such as LDR requirements.

CHAPTER 8:
Additional Tips and Advice for Accurate Waste Recognition

RECOGNIZING HAZARDOUS WASTES IS not a trivial matter. It requires applying knowledge of complex regulations, technical knowledge of the waste—often including interpretation of lab results—and sometimes knowledge of how the waste was generated. This chapter offers additional tips and advice to help ensure that your waste recognition is accurate.

Skillful Use of CAS Numbers and Material Safety Data Sheets

CAS NUMBERS

Chemicals often have various common, scientific, or brand names. To help clarify and identify chemicals, The American Chemical Society has assigned CAS numbers to tens of millions of uniquely identifiable substances including elements, minerals, mixtures, and compounds.

MATERIAL SAFETY DATA SHEETS

In the U.S., OSHA's Hazard Communication Standard requires material safety data sheets (MSDSs) for chemicals that employees may be exposed to. When properly prepared, MSDSs are a wealth of information. However, you need to be cautious about how you use them for waste characterization.

TABLE 8-1 offers guidance on using CAS numbers and MSDSs for RCRA hazardous waste characterization.

Table 8-1. RCRA Hazardous Waste Characterization: Guidance on Using CAS Numbers and MSDSs

EPA Hazardous Waste Number	CAS Numbers	MSDSs
D001–D003	**Not Helpful** ■ CAS numbers identify chemicals not characteristics.	**Helpful** ■ Helpful to identify characteristics that one should consider when evaluating wastes generated from using those raw materials. ■ However, one should not assume that characteristics of the raw materials are representative of wastes, since materials often change during use—the raw material may lose or gain characteristics during use.
D004–D011	**Not Helpful** ■ The CAS numbers listed in the regulation are for the elemental metals only. ■ The chemical species analyzed for include all chemical compounds formed by these metals, each of which has its own CAS number.	**Helpful** ■ Helpful to identify constituents in raw materials that may be in wastes. ■ However, since non-carcinogenic hazardous chemicals need only be listed on MSDSs if they are present at 1% or greater, a waste may exhibit the TC for a constituent not listed at all on an MSDS of raw materials that generate the waste.

Table 8-1. (*continued*)

EPA Hazardous Waste Number	CAS Numbers	MSDSs
D012–D043	**Helpful** ■ The CAS numbers listed in the regulation are for the actual chemical species analyzed for.	**Helpful** ■ Helpful to identify constituents in raw materials that may be in wastes. ■ However, since non-carcinogenic hazardous chemicals need only be listed on MSDSs if they are present at 1% or greater, a waste may exhibit the TC for a constituent not listed at all on an MSDS of raw materials that generate the waste.
F001–F005	**Very Helpful** ■ Chemicals used in solvents often have many synonyms.	**Very Helpful** ■ The "10% rule" applies to solvents prior to use, and MSDSs are an important source for this information.
All Other F-listed Wastes	**Limited Helpfulness** ■ Listings do not reference CAS numbers. ■ There may be situations where a CAS number can link a chemical mentioned in a listing with synonyms.	**May Be Helpful, But Be Careful** ■ MSDSs often describe how chemicals are used. So, for example, an MSDS may describe a chemical as being used for electroplating.
All K-listed Wastes	**Limited Helpfulness** ■ Listings do not reference CAS numbers. ■ There may be situations where a CAS number can link a chemical mentioned in a listing with synonyms.	**May Be Helpful, But Be Careful** ■ MSDSs often describe how chemicals are used. So, for example, an MSDS may describe a chemical as being used for electroplating.
All P-listed and U-listed Wastes	**Crucial** ■ One of the foundations of identifying P-listed and U-listed wastes is the use of CAS numbers. All chemicals listed in 40 CFR 261.33(e) and (f) have CAS numbers as part of the listings.	**Crucial** ■ MSDSs can be invaluable when identifying P-listed and U-listed wastes.

The Point When a Material Becomes a Hazardous Waste

"Point of generation" is the point in time and space when a material becomes regulated as a hazardous waste. There are two regulatory concepts that often come into play when deciding when material becomes waste: "spent material" and the 261.4(c) Rule.

SPENT MATERIAL

The definition of "spent material" in 40 CFR 261.1(c)(1) is helpful when determining the point at which a material, such as a solvent or plating bath, becomes a waste:

> "A 'spent material' is any material that has been used and as a result of contamination can no longer serve the purpose for which it was produced without processing."

Contamination that affects the usefulness of a material can be chemical or physical.

261.4(c) RULE

The regulation at 40 CFR 261.4(c) defines when a hazardous waste that is generated becomes subject to regulation as a hazardous waste.

A hazardous waste which is generated in:

➤ a product or raw material storage tank, *or*

➤ a product or raw material transport vehicle or vessel, *or*

➤ a product or raw material pipeline, *or*

➤ a manufacturing process unit or an associated non-waste-treatment-manufacturing unit;

is not subject to regulation as hazardous waste:

➤ until it exits the unit in which it was generated, unless the unit is a surface impoundment, *or*

➤ unless the hazardous waste remains in the unit more than 90 days after the unit ceases to be operated for manufacturing, or for storage or transportation of product or raw materials.

These two concepts—spent material and 261.4(c) *exclusion*—often work together. For example, once a tank of solvent becomes "spent material" as defined, it doesn't become regulated until it meets one of the two conditions listed under 40 CFR 261.4(c).

Hazardous Terms

Disposal of a "hazardous substance" is very different from disposal of a "hazardous waste." This book has discussed in depth a group of materials called "hazardous waste." This a term of art that defines a certain subclass of materials that meet certain specific regulatory conditions. There are many terms used to define many different substances for many purposes, and TABLE 8-2 contains some of them.

Table 8-2. **Hazardous Terms: Implementing Agency, Law, Regulatory Citation, and Purpose/Scope**

Hazardous Term	Implementing Agency	Law	Regulatory Citation	Purpose/Scope
Hazardous Waste	EPA	Resource Conservation and Recovery Act (RCRA)	40 CFR 261	Waste Management
Hazardous Constituent	EPA	Resource Conservation and Recovery Act (RCRA)	40 CFR 261	Waste Management
Hazardous Chemical	OSHA	Hazard Communication Standard	29 CFR 1910.1200	Chemical Exposure/ Worker Right-to-Know
Hazardous Substance	EPA	Comprehensive Environmental Response, Compensation, and Liability Act (CERCLA) aka "Superfund"	40 CFR 302	Release Reporting/ Waste Site Cleanups
Extremely Hazardous Substance	EPA	Superfund Amendments and Reauthorization Act (SARA)	40 CFR 355, Appendix A	Chemical Storage and Releases/ Community Right-to-Know
Toxic Chemical	EPA	Emergency Planning and Community Right-to-Know Act (EPCRA)	40 CFR 372	Chemical Storage and Releases/ Community Right-to-Know
Hazardous Material	DOT	Hazardous Materials Control Act	49 CFR 171	Transportation
Hazardous Air Pollutant	EPA	Clean Air Act	40 CFR Part 63	Air Pollution— Emissions from Sources
Criteria Pollutant	EPA	Clean Air Act	40 CFR Part 50	Air Pollution— Ambient Air Quality
Toxic Pollutant	EPA	Clean Water Act	40 CFR 401.15	Water Pollution
Priority Pollutants	EPA	Clean Water Act	40 CFR 423, Appendix A	Water Pollution— Toxic Pollutants with Established Test Methods

CHAPTER 8: REVIEW QUESTIONS

Questions

1. CAS numbers are useful for identifying which kind of hazardous wastes?
(Check all that apply.)

   ☐ A. discarded commercial chemical products (P-listed and U-listed wastes)

   ☐ B. every hazardous waste

   ☐ C. F-listed solvents

   ☐ D. corrosive

2. MSDSs are useful for which waste evaluations?
(Check all that apply.)

   ☐ A. determining whether a waste should be tested for corrosivity

   ☐ B. determining whether a waste is corrosive

   ☐ C. determining whether a waste is F019

   ☐ D. determining whether TCE as a spent solvent is F001 or F002

3. It's not a hazardous waste until I get the analytical results back showing whether it is or it is not a hazardous waste.

   ☐ Truth or ☐ Myth?

4. It's not a waste until I "declare" it a waste.

   ☐ Truth or ☐ Myth?

5. You generate a waste that you know poses hazards, but it doesn't exhibit any of the RCRA hazardous characteristics and doesn't meet any of the hazardous waste listings. What should you do?

   ☐ A. classify it as a characteristic or listing that is the closest fit and manage it as RCRA hazardous waste

   ☐ B. classify it as "hazardous waste, not otherwise specified, D999" and manage it as RCRA hazardous waste

   ☐ C. classify it non-RCRA hazardous and manage it as benign

   ☐ D. classify it non-RCRA hazardous and manage it consistent with its properties

Answers

1. ☑ A. CAS numbers are very useful for identifying wastes regulated as discarded commercial chemical products under 40 CFR 261.33. The P and U lists include CAS numbers to help with classification.

   ☒ B. CAS numbers are useful for identifying every hazardous waste is an incorrect statement because chemical species is not relevant to many of the hazardous waste definitions.

   ☑ C. CAS numbers are very useful for identifying wastes regulated as F-listed solvents (F001, F002, F003, F004, and F005).

   ☒ D. CAS numbers are useful for identifying corrosive hazardous wastes is an incorrect statement because these wastes are identified by testing entirely independent of chemical species.

2. ☑ A. MSDSs are very useful for determining whether a waste should be tested for corrosivity because they are very useful for identifying which tests to perform for wastes.

   ☒ B. MSDSs are useful for determining whether a waste is corrosive is an incorrect statement because they are not useful for determining properties of a chemical after it has been used.

   ☒ C. MSDSs are useful for determining whether a waste is F019 is an incorrect statement because they are not useful for determining whether the way you use a particular chemical generates a listed hazardous waste.

   ☒ D. MSDSs are useful for determining whether TCE as a spent solvent is F001 or F002 is an incorrect statement because this determination depends on how the chemical is used, not on the properties of the chemical itself.

3. ☑ Myth

   It's not a hazardous waste until I get the analytical results back showing whether it is or it is not a hazardous waste is a myth. A hazardous waste is regulated as a hazardous waste at the "point of generation."

4. ☑ Myth

   It's not a waste until I "declare" it a waste is a myth. There are very limited circumstances for the generator to simply declare when a material has become a waste.

5. ☒ A. You should classify it as a characteristic or listing that is the closest fit and manage it as RCRA hazardous waste is an incorrect statement. Forcing a waste into a hazardous waste definition can impose management requirements that don't fit and legal liability that isn't necessary.

☒ B. You should classify it as "hazardous waste, not otherwise specified, D999" and manage it as RCRA hazardous waste is an incorrect statement. There is no such RCRA classification as "hazardous waste, not otherwise specified, D999."

☒ C. You should classify it non-RCRA hazardous and manage it as benign is an incorrect statement. Just because a waste is not a RCRA hazardous waste doesn't mean that it is benign.

☑ D. You should classify it non-RCRA hazardous and manage it consistent with its properties.

Appendix 1:
RCRA Hazardous Wastes Lists

THERE ARE THREE RCRA hazardous wastes lists: *Non-specific Source Wastes, Specific Source Wastes*, and *Discarded Commercial Chemical Products*.

Non-specific Source Wastes List (§261.31)

The *Non-specific Source Wastes* list is presented thusly:
EPA Hazardous Waste Number | (EPA Hazard Codes[1]): Regulatory text.

SPENT SOLVENTS

F001 | (T):   The following spent halogenated solvents used in degreasing: tetrachloroethylene, trichloroethylene, methylene chloride, 1,1,1-trichloroethane, carbon tetrachloride, and chlorinated fluorocarbons; all spent solvent mixtures/blends used in degreasing containing, before use, a total of 10% or more (by volume) of one or more of the above halogenated solvents or those solvents listed in F002, F004, and F005; and still bottoms from the recovery of these spent solvents and spent solvent mixtures.

F002 | (T):   The following spent halogenated solvents: tetrachloroethylene, methylene chloride, trichloroethylene, 1,1,1-trichloroethane, chlorobenzene, 1,1,2-trichloro-1,2,2-trifluoroethane, ortho-dichlorobenzene, trichlorofluoromethane, and 1,1,2-trichloroethane; all spent solvent mixtures/blends containing, before use, a total of 10% or more (by volume) of one or more of the above halogenated solvents or those solvents listed in F001, F004, and F005; and still bottoms from the recovery of these spent solvents and spent solvent mixtures.

F003 | (I):   The following spent non-halogenated solvents: xylene, acetone, ethyl acetate, ethyl benzene, ethyl ether, methyl isobutyl ketone, n-butyl alcohol, cyclohexanone, and methanol; all spent solvent mixtures/blends containing, before use, only the above spent non-halogenated solvents; and all spent solvent mixtures/blends containing, before use, one or more of the above non-halogenated solvents, and a total of 10% or more (by volume) of one or more of those solvents listed in F001, F002, F004, and F005; and still bottoms from the recovery of these spent solvents and spent solvent mixtures.

F004 | (T):   The following spent non-halogenated solvents: cresols and cresylic acid, and nitrobenzene; all spent solvent mixtures/blends containing, before use, a total of 10% or more (by volume) of one or more of the above

---

1. EPA Hazard Codes: (I) Ignitable Waste, (C) Corrosive Waste, (R) Reactive Waste, (E) Toxicity Characteristic Waste, (H) Acute Hazardous Waste, and (T) Toxic Waste.

non-halogenated solvents or those solvents listed in F001, F002, and F005; and still bottoms from the recovery of these spent solvents and spent solvent mixtures.

F005 | (I,T): The following spent non-halogenated solvents: toluene, methyl ethyl ketone, carbon disulfide, isobutanol, pyridine, benzene, 2-ethoxyethanol, and 2-nitropropane; all spent solvent mixtures/blends containing, before use, a total of 10% or more (by volume) of one or more of the above non-halogenated solvents or those solvents listed in F001, F002, or F004; and still bottoms from the recovery of these spent solvents and spent solvent mixtures.

ELECTROPLATING/CONVERSION COATING

F006 | (T): Wastewater treatment sludges from electroplating operations except from the following processes: (1) sulfuric acid anodizing of aluminum; (2) tin plating on carbon steel; (3) zinc plating (segregated basis) on carbon steel; (4) aluminum or zinc-aluminum plating on carbon steel; (5) cleaning/stripping associated with tin, zinc, and aluminum plating on carbon steel; and (6) chemical etching and milling of aluminum.

F019 | (T): Wastewater treatment sludges from the chemical conversion coating of aluminum except from zirconium phosphating in aluminum can washing when such phosphating is an exclusive conversion coating process. Wastewater treatment sludges from the manufacturing of motor vehicles using a zinc phosphating process will not be subject to this listing at the point of generation if the wastes are not placed outside on the land prior to shipment to a landfill for disposal and are either: disposed in a Subtitle D municipal or industrial landfill unit that is equipped with a single clay liner and is permitted, licensed, or otherwise authorized by the state; or disposed in a landfill unit subject to, or otherwise meeting, the landfill requirements in §258.40, §264.301, or §265.301. For the purposes of this listing, motor vehicle manufacturing is defined in paragraph (b)(4)(i) of this section and (b)(4)(ii) of this section describes the recordkeeping requirements for motor vehicle manufacturing facilities.

F007 | (R,T): Spent cyanide plating bath solutions from electroplating operations.

F008 | (R,T): Plating bath residues from the bottom of plating baths from electroplating operations where cyanides are used in the process.

F009 | (R,T): Spent stripping and cleaning bath solutions from electroplating operations where cyanides are used in the process.

METAL HEAT TREATING

F010 | (R,T): Quenching bath residues from oil baths from metal heat treating operations where cyanides are used in the process.

F011 | (R,T): Spent cyanide solutions from salt bath pot cleaning from metal heat treating operations.

F012 | (T): Quenching wastewater treatment sludges from metal heat treating operations where cyanides are used in the process.

CHLORINATED ALIPHATICS, PHENOLS, AND BENZENES/DIOXIN WASTES

F020 | (H): Wastes (except wastewater and spent carbon from hydrogen chloride purification) from the production or manufacturing use (as a reactant, chemical intermediate, or component in a formulating process) of tri- or tetrachlorophenol, or of intermediates used to produce their pesticide derivatives.

[This listing does not include wastes from the production of hexachlorophene from highly purified 2,4,5-trichlorophenol.]

F021 | (H): Wastes (except wastewater and spent carbon from hydrogen chloride purification) from the production or manufacturing use (as a reactant, chemical intermediate, or component in a formulating process) of pentachlorophenol, or of intermediates used to produce its derivatives.

F022 | (H): Wastes (except wastewater and spent carbon from hydrogen chloride purification) from the manufacturing use (as a reactant, chemical intermediate, or component in a formulating process) of tetra-, penta-, or hexachlorobenzenes under alkaline conditions.

F023 | (H): Wastes (except wastewater and spent carbon from hydrogen chloride purification) from the production of materials on equipment previously used for the production or manufacturing use (as a reactant, chemical intermediate, or component in a formulating process) of tri- and tetrachlorophenols.

[This listing does not include wastes from equipment used only for the production or use of hexachlorophene from highly purified 2,4,5-trichlorophenol.]

F024 | (T): Process wastes, including but not limited to, distillation residues, heavy ends, tars, and reactor clean-out wastes, from the production of certain chlorinated aliphatic hydrocarbons by free radical catalyzed processes. These chlorinated aliphatic hydrocarbons are those having carbon chain lengths ranging from one to and including five, with varying amounts and positions of chlorine substitution.

[This listing does not include wastewaters, wastewater treatment sludges, spent catalysts, and wastes listed in §261.31 or §261.32.]

F025 | (T): Condensed light ends, spent filters and filter aids, and spent desiccant wastes from the production of certain chlorinated aliphatic hydrocarbons, by free radical catalyzed processes. These chlorinated aliphatic hydrocarbons are those having carbon chain lengths ranging from one to and including five, with varying amounts and positions of chlorine substitution.

F026 | (H): Wastes (except wastewater and spent carbon from hydrogen chloride purification) from the production of materials on equipment previously used for the manufacturing use (as a reactant, chemical intermediate, or component in a formulating process) of tetra-, penta-, or hexachlorobenzene under alkaline conditions.

F027 | (H): Discarded unused formulations containing tri-, tetra-, or pentachlorophenol or discarded unused formulations containing compounds derived from these chlorophenols.

[This listing does not include formulations containing hexachlorophene synthesized from prepurified 2,4,5-trichlorophenol as the sole component.]

F028 | (T): Residues resulting from the incineration or thermal treatment of soil contaminated with EPA Hazardous Waste Nos. F020, F021, F022, F023, F026, and F027.

WOOD PRESERVATION

F032 | (T): Wastewaters (except those that have not come into contact with process contaminants), process residuals, preservative drippage, and spent formulations from wood preserving processes generated at plants that currently use or have previously used chlorophenolic formulations (except potentially cross-contaminated wastes that have had the F032 waste code deleted in accordance with §261.35 of this chapter and where the generator does not resume or initiate use of chlorophenolic formulations).

[This listing does not include K001 bottom sediment sludge from the treatment of wastewater from wood preserving processes that use creosote and/or pentachlorophenol.]

F034 | (T): Wastewaters (except those that have not come into contact with process contaminants), process residuals, preservative drippage, and spent formulations from wood preserving processes generated at plants that use creosote formulations.

[This listing does not include K001 bottom sediment sludge from the treatment of wastewater from wood preserving processes that use creosote and/or pentachlorophenol.]

F035 | (T): Wastewaters (except those that have not come into contact with process contaminants), process residuals, preservative drippage, and spent formulations from wood preserving processes generated at plants that use inorganic preservatives containing arsenic or chromium.

[This listing does not include K001 bottom sediment sludge from the treatment of wastewater from wood preserving processes that use creosote and/or pentachlorophenol.]

PETROLEUM REFINING

F037 | (T): Petroleum refinery primary oil/water/solids separation sludge—any sludge generated from the gravitational separation of oil/water/solids during the storage or treatment of process wastewaters and oily cooling wastewaters from petroleum refineries. Such sludges include, but are not limited to, those generated in oil/water/solids separators; tanks and impoundments; ditches and other conveyances; sumps; and stormwater units receiving dry weather flow.

[Sludges generated in stormwater units that do not receive dry weather flow, sludges generated in aggressive biological treatment units as defined in §261.31(b)(2) (including sludges generated in one or more additional units after wastewaters have been treated in aggressive biological treatment units), and K051 wastes are not included in this listing.]

F038 | (T):   Petroleum refinery secondary (emulsified) oil/water/solids separation sludge—any sludge and/or float generated from the physical and/or chemical separation of oil/water/solids in process wastewaters and oily cooling wastewaters from petroleum refineries. Such wastes include, but are not limited to, all sludges and floats generated in induced air flotation (IAF) units, tanks and impoundments, and all sludges generated in DAF units.

[Sludges generated in stormwater units that do not receive dry weather flow, sludges generated in aggressive biological treatment units as defined in §261.31(b)(2) (including sludges generated in one or more additional units after wastewaters have been treated in aggressive biological treatment units), and F037, K048, and K051 wastes are not included in this listing.]

MULTISOURCE LEACHATE

F039 | (T):   Leachate (liquids that have percolated through land disposed wastes) resulting from the disposal of more than one restricted waste classified as hazardous under Subpart D of this Part. (Leachate resulting from the disposal of one or more of the following EPA hazardous wastes and no other hazardous wastes retains its EPA hazardous waste number(s): F020, F021, F022, F026, F027, and/or F028.)

Specific Source Wastes List (§261.32)

The *Specific Source Wastes* list is presented thusly:

EPA Hazardous Waste Number | (EPA Hazard Codes[2]): Regulatory text.

WOOD PRESERVATION

K001 | (T):   Bottom sediment sludge from the treatment of wastewaters from wood preserving processes that use creosote and/or pentachlorophenol.

INORGANIC PIGMENTS

K002 | (T):   Wastewater treatment sludge from the production of chrome yellow and orange pigments.

K003 | (T):   Wastewater treatment sludge from the production of molybdate orange pigments.

K004 | (T):   Wastewater treatment sludge from the production of zinc yellow pigments.

K005 | (T):   Wastewater treatment sludge from the production of chrome green pigments.

K006 | (T):   Wastewater treatment sludge from the production of chrome oxide green pigments (anhydrous and hydrated).

K007 | (T):   Wastewater treatment sludge from the production of iron blue pigments.

K008 | (T):   Oven residue from the production of chrome oxide green pigments.

ORGANIC CHEMICALS

K009 | (T):   Distillation bottoms from the production of acetaldehyde from ethylene.

---

2. EPA Hazard Codes: (I) Ignitable Waste, (C) Corrosive Waste, (R) Reactive Waste, (E) Toxicity Characteristic Waste, (H) Acute Hazardous Waste, and (T) Toxic Waste.

K010 | (T): Distillation side cuts from the production of acetaldehyde from ethylene.

K011 | (R,T): Bottom stream from the wastewater stripper in the production of acrylonitrile.

K013 | (R,T): Bottom stream from the acetonitrile column in the production of acrylonitrile.

K014 | (T): Bottoms from the acetonitrile purification column in the production of acrylonitrile.

K015 | (T): Still bottoms from the distillation of benzyl chloride.

K016 | (T): Heavy ends or distillation residues from the production of carbon tetrachloride.

K017 | (T): Heavy ends (still bottoms) from the purification column in the production of epichlorohydrin.

K018 | (T): Heavy ends from the fractionation column in ethyl chloride production.

K019 | (T): Heavy ends from the distillation of ethylene dichloride in ethylene dichloride production.

K020 | (T): Heavy ends from the distillation of vinyl chloride in vinyl chloride monomer production.

K021 | (T): Aqueous spent antimony catalyst waste from fluoromethanes production.

K022 | (T): Distillation bottom tars from the production of phenol/acetone from cumene.

K023 | (T): Distillation light ends from the production of phthalic anhydride from naphthalene.

K024 | (T): Distillation bottoms from the production of phthalic anhydride from naphthalene.

K025 | (T): Distillation bottoms from the production of nitrobenzene by the nitration of benzene.

K026 | (T): Stripping still tails from the production of methyl ethyl pyridines.

K027 | (R,T): Centrifuge and distillation residues from toluene diisocyanate production.

K028 | (T): Spent catalyst from the hydrochlorinator reactor in the production of 1,1,1-trichloroethane.

K029 | (T): Waste from the product steam stripper in the production of 1,1,1-trichloroethane.

K030 | (T): Column bottoms or heavy ends from the combined production of trichloroethylene and perchloroethylene.

K083 | (T): Distillation bottoms from aniline production.

K085 | (T): Distillation or fractionation column bottoms from the production of chlorobenzenes.

K093 | (T): Distillation light ends from the production of phthalic anhydride from ortho-xylene.

K094 | (T): Distillation bottoms from the production of phthalic anhydride from ortho-xylene.

K095 | (T): Distillation bottoms from the production of 1,1,1-trichloroethane.

K096 | (T): Heavy ends from the heavy ends column from the production of 1,1,1-trichloroethane.

K103 | (T): Process residues from aniline extraction from the production of aniline.

K104 | (T): Combined wastewater streams generated from nitrobenzene/aniline production.

K105 | (T): Separated aqueous stream from the reactor product washing step in the production of chlorobenzenes.

K107 | (C,T): Column bottoms from product separation from the production of 1,1-dimethylhydrazine (UDMH) from carboxylic acid hydrazines.

K108 | (I,T): Condensed column overheads from product separation and condensed reactor vent gases from the production of 1,1-dimethylhydrazine (UDMH) from carboxylic acid hydrazides.

K109 | (T):   Spent filter cartridges from product purification from the production of 1,1-dimethylhydrazine (UDMH) from carboxylic acid hydrazides.

K110 | (T):   Condensed column overheads from intermediate separation from the production of 1,1-dimethylhydrazine (UDMH) from carboxylic acid hydrazides.

K111 | (C,T):   Product washwaters from the production of dinitrotoluene via nitration of toluene.

K112 | (T):   Reaction by-product water from the drying column in the production of toluenediamine via hydrogenation of dinitrotoluene.

K113 | (T):   Condensed liquid light ends from the purification of toluenediamine in the production of toluenediamine via hydrogenation of dinitrotoluene.

K114 | (T):   Vicinals from the purification of toluenediamine in the production of toluenediamine via hydrogenation of dinitrotoluene.

K115 | (T):   Heavy ends from the purification of toluenediamine in the production of toluenediamine via hydrogenation of dinitrotoluene.

K116 | (T):   Organic condensate from the solvent recovery column in the production of toluene diisocyanate via phosgenation of toluenediamine.

K117 | (T):   Wastewater from the reactor vent gas scrubber in the production of ethylene dibromide via bromination of ethene.

K118 | (T):   Spent adsorbent solids from purification of ethylene dibromide in the production of ethylene dibromide via bromination of ethene.

K136 | (T):   Still bottoms from the purification of ethylene dibromide in the production of ethylene dibromide via bromination of ethene.

K149 | (T):   Distillation bottoms from the production of alpha- (or methyl-) chlorinated toluenes, ring-chlorinated toluenes, benzoyl chlorides, and compounds with mixtures of these functional groups.

[This waste does not include still bottoms from the distillation of benzyl chloride.]

K150 | (T):   Organic residuals, excluding spent carbon adsorbent, from the spent chlorine gas and hydrochloric acid recovery processes associated with the production of alpha- (or methyl-) chlorinated toluenes, ring-chlorinated toluenes, benzoyl chlorides, and compounds with mixtures of these functional groups.

K151 | (T): Wastewater treatment sludges, excluding neutralization and biological sludges, generated during treatment of wastewaters from the production of alpha- (or methyl-) chlorinated toluenes, ring-chlorinated toluenes, benzoyl chlorides, and compounds with mixtures of these functional groups.

K156 | (T): Organic waste (including heavy ends, still bottoms, light ends, spent solvents, filtrates, and decantates) from the production of carbamates and carbamoyl oximes.

[This listing does not apply to wastes generated from the manufacture of 3-iodo-2-propynyl n-butylcarbamate.]

K157 | (T): Wastewaters (including scrubber waters, condenser waters, washwaters, and separation waters) from the production of carbamates and carbamoyl oximes.

[This listing does not apply to wastes generated from the manufacture of 3-iodo-2-propynyl n-butylcarbamate.]

K158 | (T): Bag house dusts and filter/separation solids from the production of carbamates and carbamoyl oximes.

[This listing does not apply to wastes generated from the manufacture of 3-iodo-2-propynyl n-butylcarbamate.]

K159 | (T): Organics from the treatment of thiocarbamate wastes.

K161 | (R,T): Purification solids (including filtration, evaporation, and centrifugation solids), bag house dust and floor sweepings from the production of dithiocarbamate acids and their salts.

[This listing does not include K125 or K126.]

K174 | (T): Wastewater treatment sludges from the production of ethylene dichloride or vinyl chloride monomer (including sludges that result from commingled ethylene dichloride or vinyl chloride monomer wastewater and other wastewater), unless the sludges meet the following conditions: (i) they are disposed of in a Subtitle C or non-hazardous landfill licensed or permitted by the state or federal government; (ii) they are not otherwise placed on the land prior to final disposal; and (iii) the generator maintains documentation demonstrating that the waste was either disposed of in an on-site landfill or consigned to a transporter or disposal facility that provided a written commitment to dispose of the waste in an off-site landfill. Respondents in any action brought to enforce the requirements of Subtitle C must, upon a showing by the government that the respondent managed wastewater treatment sludges from the production of vinyl chloride monomer or ethylene dichloride, demonstrate that they meet the terms of the exclusion set forth above. In doing so, they must provide appropriate documentation (e.g., contracts between the generator and the landfill owner/operator, invoices documenting delivery of waste to landfill, etc.) that the terms of the exclusion were met.

K175 | (T): Wastewater treatment sludges from the production of vinyl chloride monomer using mercuric chloride catalyst in an acetylene based process.

K181 | (T): Nonwastewaters from the production of dyes and/or pigments (including nonwastewaters commingled at the point of generation with nonwastewaters from other processes) that, at the point of generation, contain mass loadings of any of the constituents identified in paragraph (c) of this section that are equal to or greater than the corresponding paragraph (c) levels, as determined on a calendar year basis. These wastes will not be hazardous if the nonwastewaters are: (i) disposed in a Subtitle D landfill unit subject to the design criteria in §258.40, (ii) disposed in a Subtitle C landfill unit subject to either §264.301 or §265.301, (iii) disposed in other Subtitle D landfill units that meet the design criteria in §258.40, §264.301, or §265.301, or (iv) treated in a combustion unit that is permitted under Subtitle C, or an onsite combustion unit that is permitted under the Clean Air Act. For the purposes of this listing, dyes and/or pigments production is defined in paragraph (b)(1) of this section. Paragraph (d) of this section describes the process for demonstrating that a facility's nonwastewaters are not K181.

[This listing does not apply to wastes that are otherwise identified as hazardous under §§261.21–261.24 and 261.31–261.33 at the point of generation. Also, the listing does not apply to wastes generated before any annual mass loading limit is met.]

INORGANIC CHEMICALS

K071 | (T): Brine purification muds from the mercury cell process in chlorine production, where separately prepurified brine is not used.

K073 | (T): Chlorinated hydrocarbon waste from the purification step of the diaphragm cell process using graphite anodes in chlorine production.

K106 | (T): Wastewater treatment sludge from the mercury cell process in chlorine production.

K176 | (E): Baghouse filters from the production of antimony oxide, including filters from the production of intermediates (e.g., antimony metal or crude antimony oxide).

K177 | (T): Slag from the production of antimony oxide that is speculatively accumulated or disposed, including slag from the production of intermediates (e.g., antimony metal or crude antimony oxide).

K178 | (T): Residues from manufacturing and manufacturing-site storage of ferric chloride from acids formed during the production of titanium dioxide using the chloride-ilmenite process.

PESTICIDES

K031 | (T): By-product salts generated in the production of MSMA and cacodylic acid.

K032 | (T): Wastewater treatment sludge from the production of chlordane.

K033 | (T): Wastewater and scrub water from the chlorination of cyclopentadiene in the production of chlordane.

K034 | (T): Filter solids from the filtration of hexachlorocyclopentadiene in the production of chlordane.

K097 | (T): Vacuum stripper discharge from the chlordane chlorinator in the production of chlordane.

K035 | (T): Wastewater treatment sludges generated in the production of creosote.

K036 | (T): Still bottoms from toluene reclamation distillation in the production of disulfoton.

K037 | (T): Wastewater treatment sludges from the production of disulfoton.

K038 | (T): Wastewater from the washing and stripping of phorate production.

K039 | (T): Filter cake from the filtration of diethylphosphorodithioic acid in the production of phorate.

K040 | (T): Wastewater treatment sludge from the production of phorate.

K041 | (T): Wastewater treatment sludge from the production of toxaphene.

K098 | (T): Untreated process wastewater from the production of toxaphene.

K042 | (T): Heavy ends or distillation residues from the distillation of tetrachlorobenzene in the production of 2,4,5-T.

K043 | (T): 2,6-Dichlorophenol waste from the production of 2,4-D.

K099 | (T): Untreated wastewater from the production of 2,4-D.

K123 | (T): Process wastewater (including supernates, filtrates, and washwaters) from the production of ethylenebisdithiocarbamic acid and its salts.

K124 | (C,T): Reactor vent scrubber water from the production of ethylenebisdithiocarbamic acid and its salts.

K125 | (T): Filtration, evaporation, and centrifugation solids from the production of ethylenebisdithiocarbamic acid and its salts.

K126 | (T): Baghouse catch and floor sweepings in milling and packaging operations from the production or formulation of ethylenebisdithiocarbamic acid and its salts.

K131 | (C,T): Wastewater from the reactor and spent sulfuric acid from the acid dryer from the production of methyl bromide.

K132 | (T): Spent absorbent and wastewater separator solids from the production of methyl bromide.

EXPLOSIVES

K044 | (R): Wastewater treatment sludges from the manufacturing and processing of explosives.

K045 | (R): Spent carbon from the treatment of wastewater containing explosives.

K046 | (T): Wastewater treatment sludges from the manufacturing, formulation, and loading of lead-based initiating compounds.

K047 | (R): Pink/red water from TNT operations.

PETROLEUM REFINING

K048 | (T): Dissolved air flotation (DAF) float from the petroleum refining industry.

K049 | (T): Slop oil emulsion solids from the petroleum refining industry.

K050 | (T): Heat exchanger bundle cleaning sludge from the petroleum refining industry.

K051 | (T): API separator sludge from the petroleum refining industry.

K052 | (T): Tank bottoms (leaded) from the petroleum refining industry.

K169 | (T): Crude oil storage tank sediment from petroleum refining operations.

K170 | (T): Clarified slurry oil tank sediment and/or in-line filter/separation solids from petroleum refining operations.

K171 | (I,T): Spent hydrotreating catalyst from petroleum refining operations, including guard beds used to desulfurize feeds to other catalytic reactors.

[This listing does not include inert support media.]

K172 | (I,T): Spent hydrorefining catalyst from petroleum refining operations, including guard beds used to desulfurize feeds to other catalytic reactors.

[This listing does not include inert support media.]

IRON AND STEEL

K061 | (T): Emission control dust/sludge from the primary production of steel in electric furnaces.

K062 | (C,T): Spent pickle liquor generated by steel finishing operations of facilities within the iron and steel industry (SIC Codes 331 and 332).

PRIMARY COPPER

K064 | (T):   Acid plant blowdown slurry/sludge resulting from the thickening of blowdown slurry from primary copper production.

PRIMARY LEAD

K065 | (T):   Surface impoundment solids contained in and dredged from surface impoundments at primary lead smelting facilities.

PRIMARY ZINC

K066 | (T):   Sludge from treatment of process wastewater and/or acid plant blowdown from primary zinc production.

PRIMARY ALUMINUM

K088 | (T):   Spent potliners from primary aluminum reduction.

FERROALLOYS

K090 | (T):   Emission control dust or sludge from ferrochromiumsilicon production.

K091 | (T):   Emission control dust or sludge from ferrochromium production.

SECONDARY LEAD

K069 | (T):   Emission control dust/sludge from secondary lead smelting.

[Note: This listing is stayed administratively for sludge generated from secondary acid scrubber systems. The stay will remain in effect until further administrative action is taken. If EPA takes further action effecting this stay, EPA will publish a notice of the action in the FR.]

K100 | (T):   Waste leaching solution from acid leaching of emission control dust/sludge from secondary lead smelting.

VETERINARY PHARMACEUTICALS

K084 | (T): Wastewater treatment sludges generated during the production of veterinary pharmaceuticals from arsenic or organo-arsenic compounds.

K101 | (T): Distillation tar residues from the distillation of aniline-based compounds in the production of veterinary pharmaceuticals from arsenic or organo-arsenic compounds.

K102 | (T): Residue from the use of activated carbon for decolorization in the production of veterinary pharmaceuticals from arsenic or organo-arsenic compounds.

INK FORMULATION

K086 | (T): Solvent washes and sludges, caustic washes and sludges, or water washes and sludges from cleaning tubs and equipment used in the formulation of ink from pigments, driers, soaps, and stabilizers containing chromium and lead.

COKING

K060 | (T): Ammonia still lime sludge from coking operations.

K087 | (T): Decanter tank tar sludge from coking operations.

K141 | (T): Process residues from the recovery of coal tar, including, but not limited to, collecting sump residues from the production of coke from coal or the recovery of coke by-products produced from the coal.

[This listing does not include K087 (decanter tank tar sludges from coking operations).]

K142 | (T): Tar storage tank residues from the production of coke from coal or from the recovery of coke by-products from coal.

K143 | (T): Process residues from the recovery of light oil, including, but not limited to, those generated in stills, decanters, and wash oil recovery units from the recovery of coke by-products produced from coal.

K144 | (T): Wastewater sump residues from light oil refining, including, but not limited to, intercepting or contamination sump sludges from the recovery of coke by-products produced from coal.

K145 | (T): Residues from napthalene collection and recovery operations from the recovery of coke by-products produced from coal.

K147 | (T): Tar storage tank residues from coal tar refining.

K148 | (T): Residues from coal tar distillation, including but not limited to, still bottoms.

Discarded Commercial Chemical Products List [§261.33(e)–(f)]

The *Discarded Commercial Chemical Products* list is comprised of P-listed waste and U-listed waste. P-listed waste is acute hazardous in list §261.33(e) and assigned EPA Hazardous Waste Numbers of a "P" followed by a three-digit number. U-listed waste is hazardous in list §261.33(f) and assigned EPA Hazardous Waste Numbers of a "U" followed by a three-digit number.

The *Discarded Commercial Chemical Products* list is presented in two tables:

■ Table Appendix 1-1 contains P-listed waste (see page 128) and U-listed waste (see page 133) in *alphabetical order by substance*.

■ Table Appendix 1-2 contains P-listed waste (see page 143) and U-listed waste (see page 148) in *alphanumeric order by EPA Hazardous Waste Number*.

Notes for Table Appendix 1-1 and Table Appendix 1-2 are as follows:

*CAS Number: Given for parent compound only.

NA: Not applicable.

P-listed Waste: The primary hazardous properties of these materials is indicated by the letters T (toxicity) and R (reactivity). Absence of a letter indicates that the compound only is listed for acute toxicity.

U-listed Waste: The primary hazardous properties of these materials is indicated by the letters T (toxicity), R (reactivity), I (ignitability), and C (corrosivity). Absence of a letter indicates that the compound is only listed for toxicity.

Source: http://ecfr.gpoaccess.gov/cgi/t/text/text-idx?c=ecfr&sid=f3fc1aeedcb0cc87bf4367 f0a9cc03d0&rgn=div5&view=text&node=40:26.0.1.1.2&idno=40#40:26.0.1.1.2.4.1.4

Table Appendix 1-1. Discarded Commercial Chemical Products (P & U Lists): Alphabetical by Substance

EPA Haz. Waste #	CAS* Number	Substance	EPA Haz. Waste #	CAS* Number	Substance
— Acutely Hazardous §261.33(e) —			P077	100-01-6	Benzenamine, 4-nitro-
P023	107-20-0	Acetaldehyde, chloro-	P028	100-44-7	Benzene, (chloromethyl)-
P002	591-08-2	Acetamide, N-(aminothioxomethyl)-	P042	51-43-4	1,2-Benzenediol, 4-[1-hydroxy-2-(methylamino)ethyl]-, (R)-
P057	640-19-7	Acetamide, 2-fluoro-	P046	122-09-8	Benzeneethanamine, alpha,alpha-dimethyl-
P058	62-74-8	Acetic acid, fluoro-, sodium salt			
P002	591-08-2	1-Acetyl-2-thiourea	P014	108-98-5	Benzenethiol
P003	107-02-8	Acrolein	P127	1563-66-2	7-Benzofuranol, 2,3-dihydro-2,2-dimethyl-, methylcarbamate
P070	116-06-3	Aldicarb	P188	57-64-7	Benzoic acid, 2-hydroxy-, compd. with (3aS-cis)-1,2,3,3a,8,8a-hexahydro-1,3a,8-trimethylpyrrolo[2,3-b]indol-5-yl methylcarbamate ester (1:1)
P203	1646-88-4	Aldicarb sulfone			
P004	309-00-2	Aldrin			
P005	107-18-6	Allyl alcohol			
P006	20859-73-8	Aluminum phosphide (R,T)	P001	81-81-2*	2H-1-Benzopyran-2-one, 4-hydroxy-3-(3-oxo-1-phenylbutyl)-, & salts, when present at concentrations greater than 0.3%
P007	2763-96-4	5-(Aminomethyl)-3-isoxazolol			
P008	504-24-5	4-Aminopyridine	P028	100-44-7	Benzyl chloride
P009	131-74-8	Ammonium picrate (R)	P015	7440-41-7	Beryllium powder
P119	7803-55-6	Ammonium vanadate	P017	598-31-2	Bromoacetone
P099	506-61-6	Argentate(1-), bis(cyano-C)-, potassium	P018	357-57-3	Brucine
P010	7778-39-4	Arsenic acid H3AsO4	P045	39196-18-4	2-Butanone, 3,3-dimethyl-1-(methylthio)-, O-[(methylamino)carbonyl] oxime
P012	1327-53-3	Arsenic oxide As2O3			
P011	1303-28-2	Arsenic oxide As2O5	P021	592-01-8	Calcium cyanide
P011	1303-28-2	Arsenic pentoxide	P021	592-01-8	Calcium cyanide Ca(CN)$_2$
P012	1327-53-3	Arsenic trioxide	P189	55285-14-8	Carbamic acid, [(dibutylamino)- thio]methyl-, 2,3-dihydro-2,2-dimethyl-7-benzofuranyl ester
P038	692-42-2	Arsine, diethyl-			
P036	696-28-6	Arsonous dichloride, phenyl-	P191	644-64-4	Carbamic acid, dimethyl-, 1-[(dimethyl-amino)carbonyl]-5-methyl-1H-pyrazol-3-yl ester
P054	151-56-4	Aziridine			
P067	75-55-8	Aziridine, 2-methyl-	P192	119-38-0	Carbamic acid, dimethyl-, 3-methyl-1-(1-methylethyl)-1H-pyrazol-5-yl ester
P013	542-62-1	Barium cyanide	P190	1129-41-5	Carbamic acid, methyl-, 3-methylphenyl ester
P024	106-47-8	Benzenamine, 4-chloro-			

Table Appendix 1-1. (*continued*)

EPA Haz. Waste #	CAS* Number	Substance	EPA Haz. Waste #	CAS* Number	Substance
P127	1563-66-2	Carbofuran	P060	465-73-6	1,4,5,8-Dimethanonaphthalene, 1,2,3,4,10,10-hexachloro-1,4,4a,5,8,8a-hexahydro-, (1alpha, 4alpha,4abeta,5beta,8beta,8abeta)-
P022	75-15-0	Carbon disulfide
P095	75-44-5	Carbonic dichloride
P189	55285-14-8	Carbosulfan	P037	60-57-1	2,7:3,6-Dimethanonaphth[2,3-b] oxirene, 3,4,5,6,9,9-hexachloro-1a,2,2a,3,6,6a,7,7a-octahydro-, (1aalpha,2beta,2aalpha,3beta,6beta, 6aalpha,7beta,7aalpha)-
P023	107-20-0	Chloroacetaldehyde
P024	106-47-8	p-Chloroaniline
P026	5344-82-1	1-(o-Chlorophenyl)thiourea	P051	72-20-8*	2,7:3,6-Dimethanonaphth[2,3-b] oxirene, 3,4,5,6,9,9-hexachloro-1a,2,2a,3,6,6a,7,7a-octahydro-, (1aalpha,2beta,2abeta,3alpha,6alpha, 6abeta,7beta,7aalpha)-, & metabolites
P027	542-76-7	3-Chloropropionitrile
P029	544-92-3	Copper cyanide
P029	544-92-3	Copper cyanide Cu(CN)	P044	60-51-5	Dimethoate
P202	64-00-6	m-Cumenyl methylcarbamate	P046	122-09-8	alpha,alpha-Dimethylphenethylamine
P030	NA	Cyanides (soluble cyanide salts), not otherwise specified	P191	644-64-4	Dimetilan
P031	460-19-5	Cyanogen	P047	534-52-1*	4,6-Dinitro-o-cresol, & salts
P033	506-77-4	Cyanogen chloride	P048	51-28-5	2,4-Dinitrophenol
P033	506-77-4	Cyanogen chloride (CN)Cl	P020	88-85-7	Dinoseb
P034	131-89-5	2-Cyclohexyl-4,6-dinitrophenol	P085	152-16-9	Diphosphoramide, octamethyl-
P016	542-88-1	Dichloromethyl ether	P111	107-49-3	Diphosphoric acid, tetraethyl ester
P036	696-28-6	Dichlorophenylarsine	P039	298-04-4	Disulfoton
P037	60-57-1	Dieldrin	P049	541-53-7	Dithiobiuret
P038	692-42-2	Diethylarsine	P185	26419-73-8	1,3-Dithiolane-2-carboxaldehyde, 2,4-dimethyl-, O-[(methylamino)-carbonyl]oxime
P041	311-45-5	Diethyl-p-nitrophenyl phosphate
P040	297-97-2	O,O-Diethyl O-pyrazinyl phosphorothioate	P050	115-29-7	Endosulfan
P043	55-91-4	Diisopropylfluorophosphate (DFP)	P088	145-73-3	Endothall
P004	309-00-2	1,4,5,8-Dimethanonaphthalene, 1,2,3,4,10,10-hexachloro-1,4,4a,5,8,8a-hexahydro-, (1alpha, 4alpha,4abeta,5alpha,8alpha,8abeta)-	P051	72-20-8	Endrin
			P051	72-20-8	Endrin & metabolites
			P042	51-43-4	Epinephrine
			P031	460-19-5	Ethanedinitrile

Table Appendix 1-1. (*continued*)

EPA Haz. Waste #	CAS* Number	Substance
P194	23135-22-0	Ethanimidothioic acid, 2-(dimethylamino)-N-[[(methylamino) carbonyl]oxy]-2-oxo-, methyl ester
P066	16752-77-5	Ethanimidothioic acid, N-[[(methyl amino)carbonyl]oxy]-, methyl ester
P101	107-12-0	Ethyl cyanide
P054	151-56-4	Ethyleneimine
P097	52-85-7	Famphur
P056	7782-41-4	Fluorine
P057	640-19-7	Fluoroacetamide
P058	62-74-8	Fluoroacetic acid, sodium salt
P198	23422-53-9	Formetanate hydrochloride
P197	17702-57-7	Formparanate
P065	628-86-4	Fulminic acid, mercury(2+) salt (R,T)
P059	76-44-8	Heptachlor
P062	757-58-4	Hexaethyl tetraphosphate
P116	79-19-6	Hydrazinecarbothioamide
P068	60-34-4	Hydrazine, methyl-
P063	74-90-8	Hydrocyanic acid
P063	74-90-8	Hydrogen cyanide
P096	7803-51-2	Hydrogen phosphide
P060	465-73-6	Isodrin
P192	119-38-0	Isolan
P202	64-00-6	3-Isopropylphenyl N-methylcarbamate
P007	2763-96-4	3(2H)-Isoxazolone, 5-(aminomethyl)-
P196	15339-36-3	Manganese, bis(dimethylcarbamodithioato-S,S´)-,
P196	15339-36-3	Manganese dimethyldithiocarbamate
P092	62-38-4	Mercury, (acetato-O)phenyl-
P065	628-86-4	Mercury fulminate (R,T)

EPA Haz. Waste #	CAS* Number	Substance
P082	62-75-9	Methanamine, N-methyl-N-nitroso-
P064	624-83-9	Methane, isocyanato-
P016	542-88-1	Methane, oxybis[chloro-
P112	509-14-8	Methane, tetranitro- (R)
P118	75-70-7	Methanethiol, trichloro-
P198	23422-53-9	Methanimidamide, N,N-dimethyl-N´-[3-[[(methylamino)-carbonyl]oxy] phenyl]-, monohydrochloride
P197	17702-57-7	Methanimidamide, N,N-dimethyl-N´-[2-methyl-4-[[(methylamino)carbonyl] oxy]phenyl]-
P050	115-29-7	6,9-Methano-2,4,3-benzodioxathiepin, 6,7,8,9,10,10-hexachloro-1,5,5a,6,9,9a-hexahydro-, 3-oxide
P059	76-44-8	4,7-Methano-1H-indene, 1,4,5,6,7,8, 8-heptachloro-3a,4,7,7a-tetrahydro-
P199	2032-65-7	Methiocarb
P066	16752-77-5	Methomyl
P068	60-34-4	Methyl hydrazine
P064	624-83-9	Methyl isocyanate
P069	75-86-5	2-Methyllactonitrile
P071	298-00-0	Methyl parathion
P190	1129-41-5	Metolcarb
P128	315-8-4	Mexacarbate
P072	86-88-4	alpha-Naphthylthiourea
P073	13463-39-3	Nickel carbonyl
P073	13463-39-3	Nickel carbonyl Ni(CO)₄, (T-4)-
P074	557-19-7	Nickel cyanide
P074	557-19-7	Nickel cyanide Ni(CN)₂
P075	54-11-5*	Nicotine, & salts
P076	10102-43-9	Nitric oxide

Table Appendix 1-1. (*continued*)

EPA Haz. Waste #	CAS* Number	Substance	EPA Haz. Waste #	CAS* Number	Substance
P077	100-01-6	p-Nitroaniline	P094	298-02-2	Phorate
P078	10102-44-0	Nitrogen dioxide	P095	75-44-5	Phosgene
P076	10102-43-9	Nitrogen oxide NO	P096	7803-51-2	Phosphine
P078	10102-44-0	Nitrogen oxide NO$_2$	P041	311-45-5	Phosphoric acid, diethyl 4-nitrophenyl ester
P081	55-63-0	Nitroglycerine (R)	P039	298-04-4	Phosphorodithioic acid, O,O-diethyl S-[2-(ethylthio)ethyl] ester
P082	62-75-9	N-Nitrosodimethylamine			
P084	4549-40-0	N-Nitrosomethylvinylamine	P094	298-02-2	Phosphorodithioic acid, O,O-diethyl S-[(ethylthio)methyl] ester
P085	152-16-9	Octamethylpyrophosphoramide			
P087	20816-12-0	Osmium oxide OsO$_4$, (T-4)-	P044	60-51-5	Phosphorodithioic acid, O,O-dimethyl S-[2-(methylamino)-2-oxoethyl] ester
P087	20816-12-0	Osmium tetroxide	P043	55-91-4	Phosphorofluoridic acid, bis(1-methylethyl) ester
P088	145-73-3	7-Oxabicyclo[2.2.1]heptane-2,3-dicarboxylic acid	P089	56-38-2	Phosphorothioic acid, O,O-diethyl O-(4-nitrophenyl) ester
P194	23135-22-0	Oxamyl			
P089	56-38-2	Parathion	P040	297-97-2	Phosphorothioic acid, O,O-diethyl O-pyrazinyl ester
P034	131-89-5	Phenol, 2-cyclohexyl-4,6-dinitro-	P097	52-85-7	Phosphorothioic acid, O-[4-[(dimethylamino)sulfonyl]phenyl] O,O-dimethyl ester
P048	51-28-5	Phenol, 2,4-dinitro-			
P047	534-52-1*	Phenol, 2-methyl-4,6-dinitro-, & salts			
P020	88-85-7	Phenol, 2-(1-methylpropyl)-4,6-dinitro-	P071	298-00-0	Phosphorothioic acid, O,O,-dimethyl O-(4-nitrophenyl) ester
P009	131-74-8	Phenol, 2,4,6-trinitro-, ammonium salt (R)	P204	57-47-6	Physostigmine
			P188	57-64-7	Physostigmine salicylate
P128	315-18-4	Phenol, 4-(dimethylamino)-3,5-dimethyl-, methylcarbamate (ester)	P110	78-00-2	Plumbane, tetraethyl-
			P098	151-50-8	Potassium cyanide
P199	2032-65-7	Phenol, (3,5-dimethyl-4-(methylthio)-, methylcarbamate	P098	151-50-8	Potassium cyanide K(CN)
P202	64-00-6	Phenol, 3-(1-methylethyl)-, methyl carbamate	P099	506-61-6	Potassium silver cyanide
			P201	2631-37-0	Promecarb
P201	2631-37-0	Phenol, 3-methyl-5-(1-methylethyl)-, methyl carbamate	P070	116-06-3	Propanal, 2-methyl-2-(methylthio)-, O-[(methylamino)carbonyl]oxime
P092	62-38-4	Phenylmercury acetate	P203	1646-88-4	Propanal, 2-methyl-2-(methyl-sulfonyl)-, O-[(methylamino)carbonyl]oxime
P093	103-85-5	Phenylthiourea			

Table Appendix 1-1. (*continued*)

EPA Haz. Waste #	CAS* Number	Substance	EPA Haz. Waste #	CAS* Number	Substance
P101	107-12-0	Propanenitrile	P111	107-49-3	Tetraethyl pyrophosphate
P027	542-76-7	Propanenitrile, 3-chloro-	P112	509-14-8	Tetranitromethane (R)
P069	75-86-5	Propanenitrile, 2-hydroxy-2-methyl-	P062	757-58-4	Tetraphosphoric acid, hexaethyl ester
P081	55-63-0	1,2,3-Propanetriol, trinitrate (R)	P113	1314-32-5	Thallic oxide
P017	598-31-2	2-Propanone, 1-bromo-	P113	1314-32-5	Thallium oxide Tl_2O_3
P102	107-19-7	Propargyl alcohol	P114	12039-52-0	Thallium(I) selenite
P003	107-02-8	2-Propenal	P115	7446-18-6	Thallium(I) sulfate
P005	107-18-6	2-Propen-1-ol	P109	3689-24-5	Thiodiphosphoric acid, tetraethyl ester
P067	75-55-8	1,2-Propylenimine	P045	39196-18-4	Thiofanox
P102	107-19-7	2-Propyn-1-ol	P049	541-53-7	Thioimidodicarbonic diamide [(H_2N)C(S)]$_2$NH
P008	504-24-5	4-Pyridinamine			
P075	54-11-5*	Pyridine, 3-(1-methyl-2-pyrrolidinyl)-, (S)-, & salts	P014	108-98-5	Thiophenol
			P116	79-19-6	Thiosemicarbazide
P204	57-47-6	Pyrrolo[2,3-b]indol-5-ol, 1,2,3,3a,8, 8a-hexahydro-1,3a,8-trimethyl-, methylcarbamate (ester), (3aS-cis)-	P026	5344-82-1	Thiourea, (2-chlorophenyl)-
			P072	86-88-4	Thiourea, 1-naphthalenyl-
P114	12039-52-0	Selenious acid, dithallium(1+) salt	P093	103-85-5	Thiourea, phenyl-
P103	630-10-4	Selenourea	P185	26419-73-8	Tirpate
P104	506-64-9	Silver cyanide	P123	8001-35-2	Toxaphene
P104	506-64-9	Silver cyanide Ag(CN)	P118	75-70-7	Trichloromethanethiol
P105	26628-22-8	Sodium azide	P119	7803-55-6	Vanadic acid, ammonium salt
P106	143-33-9	Sodium cyanide	P120	1314-62-1	Vanadium oxide V_2O_5
P106	143-33-9	Sodium cyanide Na(CN)	P120	1314-62-1	Vanadium pentoxide
P108	57-24-9*	Strychnidin-10-one, & salts	P084	4549-40-0	Vinylamine, N-methyl-N-nitroso-
P018	357-57-3	Strychnidin-10-one, 2,3-dimethoxy-	P001	81-81-2*	Warfarin, & salts, when present at concentrations greater than 0.3%
P108	57-24-9*	Strychnine, & salts			
P115	7446-18-6	Sulfuric acid, dithallium(1+) salt	P205	137-30-4	Zinc, bis(dimethylcarbamodithioato-S,S´)-,
P109	3689-24-5	Tetraethyldithiopyrophosphate	P121	557-21-1	Zinc cyanide
P110	78-00-2	Tetraethyl lead	P121	557-21-1	Zinc cyanide Zn(CN)$_2$

Table Appendix 1-1. (*continued*)

EPA Haz. Waste #	CAS* Number	Substance
P122	1314-84-7	Zinc phosphide Zn_3P_2, when present at concentrations greater than 10% (R,T)
P205	137-30-4	Ziram
— Hazardous §261.33(f) —		
U394	30558-43-1	A2213
U001	75-07-0	Acetaldehyde (I)
U034	75-87-6	Acetaldehyde, trichloro-
U187	62-44-2	Acetamide, N-(4-ethoxyphenyl)-
U005	53-96-3	Acetamide, N-9H-fluoren-2-yl-
U240	94-75-7*	Acetic acid, (2,4-dichlorophenoxy)-, salts & esters
U112	141-78-6	Acetic acid, ethyl ester (I)
U144	301-04-2	Acetic acid, lead(2+) salt
U214	563-68-8	Acetic acid, thallium(1+) salt
See F027	93-76-5	Acetic acid, (2,4,5-trichlorophenoxy)-
U002	67-64-1	Acetone (I)
U003	75-05-8	Acetonitrile (I,T)
U004	98-86-2	Acetophenone
U005	53-96-3	2-Acetylaminofluorene
U006	75-36-5	Acetyl chloride (C,R,T)
U007	79-06-1	Acrylamide
U008	79-10-7	Acrylic acid (I)
U009	107-13-1	Acrylonitrile
U011	61-82-5	Amitrole
U012	62-53-3	Aniline (I,T)
U136	75-60-5	Arsinic acid, dimethyl-
U014	492-80-8	Auramine
U015	115-02-6	Azaserine

EPA Haz. Waste #	CAS* Number	Substance
U010	50-07-7	Azirino[2′,3′:3,4]pyrrolo[1,2-a]indole-4,7-dione, 6-amino-8-[[(amino carbonyl)oxy]methyl]-1,1a,2,8,8a,8b-hexahydro-8a-methoxy-5-methyl-, [1aS-(1aalpha,8beta,8aalpha,8balpha)]-
U280	101-27-9	Barban
U278	22781-23-3	Bendiocarb
U364	22961-82-6	Bendiocarb phenol
U271	17804-35-2	Benomyl
U157	56-49-5	Benz[j]aceanthrylene, 1,2-dihydro-3-methyl-
U016	225-51-4	Benz[c]acridine
U017	98-87-3	Benzal chloride
U192	23950-58-5	Benzamide, 3,5-dichloro-N-(1,1-dimethyl-2-propynyl)-
U018	56-55-3	Benz[a]anthracene
U094	57-97-6	Benz[a]anthracene, 7,12-dimethyl-
U012	62-53-3	Benzenamine (I,T)
U014	492-80-8	Benzenamine, 4,4′,-carbonimidoylbis[N,N-dimethyl-
U049	3165-93-3	Benzenamine, 4-chloro-2-methyl-, hydrochloride
U093	60-11-7	Benzenamine, N,N-dimethyl-4-(phenylazo)-
U328	95-53-4	Benzenamine, 2-methyl-
U353	106-49-0	Benzenamine, 4-methyl-
U158	101-14-4	Benzenamine, 4,4′-methylenebis[2-chloro-
U222	636-21-5	Benzenamine, 2-methyl-, hydrochloride
U181	99-55-8	Benzenamine, 2-methyl-5-nitro-
U019	71-43-2	Benzene (I,T)

Table Appendix 1-1. (*continued*)

EPA Haz. Waste #	CAS* Number	Substance
U038	510-15-6	Benzeneacetic acid, 4-chloro-alpha-(4-chlorophenyl)-alpha-hydroxy-, ethyl ester
U030	101-55-3	Benzene, 1-bromo-4-phenoxy-
U035	305-03-3	Benzenebutanoic acid, 4-[bis(2-chloroethyl)amino]-
U037	108-90-7	Benzene, chloro-
U221	25376-45-8	Benzenediamine, ar-methyl-
U028	117-81-7	1,2-Benzenedicarboxylic acid, bis(2-ethylhexyl) ester
U069	84-74-2	1,2-Benzenedicarboxylic acid, dibutyl ester
U088	84-66-2	1,2-Benzenedicarboxylic acid, diethyl ester
U102	131-11-3	1,2-Benzenedicarboxylic acid, dimethyl ester
U107	117-84-0	1,2-Benzenedicarboxylic acid, dioctyl ester
U070	95-50-1	Benzene, 1,2-dichloro-
U071	541-73-1	Benzene, 1,3-dichloro-
U072	106-46-7	Benzene, 1,4-dichloro-
U060	72-54-8	Benzene, 1,1´-(2,2-dichloroethylidene) bis[4-chloro-
U017	98-87-3	Benzene, (dichloromethyl)-
U223	26471-62-5	Benzene, 1,3,-diisocyanatomethyl- (R,T)
U239	1330-20-7	Benzene, dimethyl- (I)
U201	108-46-3	1,3-Benzenediol
U127	118-74-1	Benzene, hexachloro-
U056	110-82-7	Benzene, hexahydro- (I)
U220	108-88-3	Benzene, methyl-
U105	121-14-2	Benzene, 1-methyl-2,4-dinitro-

EPA Haz. Waste #	CAS* Number	Substance
U106	606-20-2	Benzene, 2-methyl-1,3-dinitro-
U055	98-82-8	Benzene, (1-methylethyl)- (I)
U169	98-95-3	Benzene, nitro-
U183	608-93-5	Benzene, pentachloro-
U185	82-68-8	Benzene, pentachloronitro-
U020	98-09-9	Benzenesulfonic acid chloride (C,R)
U020	98-09-9	Benzenesulfonyl chloride (C,R)
U207	95-94-3	Benzene, 1,2,4,5-tetrachloro-
U061	50-29-3	Benzene, 1,1´-(2,2,2-trichloro ethylidene)bis[4-chloro-
U247	72-43-5	Benzene, 1,1´-(2,2,2-trichloro ethylidene)bis[4-methoxy-
U023	98-07-7	Benzene, (trichloromethyl)-
U234	99-35-4	Benzene, 1,3,5-trinitro-
U021	92-87-5	Benzidine
U278	22781-23-3	1,3-Benzodioxol-4-ol, 2,2-dimethyl-, methyl carbamate
U364	22961-82-6	1,3-Benzodioxol-4-ol, 2,2-dimethyl-,
U203	94-59-7	1,3-Benzodioxole, 5-(2-propenyl)-
U141	120-58-1	1,3-Benzodioxole, 5-(1-propenyl)-
U367	1563-38-8	7-Benzofuranol, 2,3-dihydro-2,2-dimethyl-
U090	94-58-6	1,3-Benzodioxole, 5-propyl-
U064	189-55-9	Benzo[rst]pentaphene
U248	81-81-2*	2H-1-Benzopyran-2-one, 4-hydroxy-3-(3-oxo-1-phenyl-butyl)-, & salts, when present at concentrations of 0.3% or less
U022	50-32-8	Benzo[a]pyrene
U197	106-51-4	p-Benzoquinone
U023	98-07-7	Benzotrichloride (C,R,T)

Table Appendix 1-1. (*continued*)

EPA Haz. Waste #	CAS* Number	Substance	EPA Haz. Waste #	CAS* Number	Substance
U085	1464-53-5	2,2'-Bioxirane	U178	615-53-2	Carbamic acid, methylnitroso-, ethyl ester
U021	92-87-5	[1,1'-Biphenyl]-4,4'-diamine	U373	122-42-9	Carbamic acid, phenyl-, 1-methylethyl ester
U073	91-94-1	[1,1'-Biphenyl]-4,4'-diamine, 3,3'-dichloro-	U409	23564-05-8	Carbamic acid, [1,2-phenylenebis (iminocarbonothioyl)]bis-, dimethyl ester
U091	119-90-4	[1,1'-Biphenyl]-4,4'-diamine, 3,3'-dimethoxy-	U097	79-44-7	Carbamic chloride, dimethyl-
U095	119-93-7	[1,1'-Biphenyl]-4,4'-diamine, 3,3'-dimethyl-	U389	2303-17-5	Carbamothioic acid, bis(1-methylethyl)-, S-(2,3,3-trichloro-2-propenyl) ester
U225	75-25-2	Bromoform	U387	52888-80-9	Carbamothioic acid, dipropyl-, S-(phenylmethyl) ester
U030	101-55-3	4-Bromophenyl phenyl ether			
U128	87-68-3	1,3-Butadiene, 1,1,2,3,4,4-hexachloro-	U114	111-54-6*	Carbamodithioic acid, 1,2-ethanediylbis-, salts & esters
U172	924-16-3	1-Butanamine, N-butyl-N-nitroso-	U062	2303-16-4	Carbamodithioic acid, bis(1-methylethyl)-, S-(2,3-dichloro-2-propenyl) ester
U031	71-36-3	1-Butanol (I)			
U159	78-93-3	2-Butanone (I,T)			
U160	1338-23-4	2-Butanone, peroxide (R,T)	U279	63-25-2	Carbaryl
U053	4170-30-3	2-Butenal	U372	10605-21-7	Carbendazim
U074	764-41-0	2-Butene, 1,4-dichloro- (I,T)	U367	1563-38-8	Carbofuran phenol
U143	303-34-4	2-Butenoic acid, 2-methyl-, 7-[[2,3-dihydroxy-2-(1-methoxyethyl)-3-methyl-1-oxobutoxy]methyl]-2,3,5,7a-tetrahydro-1H-pyrrolizin-1-yl ester, [1S-[1alpha(Z),7(2S*,3R*),7aalpha]]-	U215	6533-73-9	Carbonic acid, dithallium(1+) salt
			U033	353-50-4	Carbonic difluoride
			U156	79-22-1	Carbonochloridic acid, methyl ester (I,T)
U031	71-36-3	n-Butyl alcohol (I)	U033	353-50-4	Carbon oxyfluoride (R,T)
U136	75-60-5	Cacodylic acid	U211	56-23-5	Carbon tetrachloride
U032	13765-19-0	Calcium chromate	U034	75-87-6	Chloral
U372	10605-21-7	Carbamic acid, 1H-benzimidazol-2-yl, methyl ester	U035	305-03-3	Chlorambucil
U271	17804-35-2	Carbamic acid, [1-[(butylamino) carbonyl]-1H-benzimidazol-2-yl]-, methyl ester	U036	57-74-9	Chlordane, alpha & gamma isomers
			U026	494-03-1	Chlornaphazin
U280	101-27-9	Carbamic acid, (3-chlorophenyl)-, 4-chloro-2-butynyl ester	U037	108-90-7	Chlorobenzene
U238	51-79-6	Carbamic acid, ethyl ester	U038	510-15-6	Chlorobenzilate

Table Appendix 1-1. (*continued*)

EPA Haz. Waste #	CAS* Number	Substance	EPA Haz. Waste #	CAS* Number	Substance
U039	59-50-7	p-Chloro-m-cresol	U064	189-55-9	Dibenzo[a,i]pyrene
U042	110-75-8	2-Chloroethyl vinyl ether	U066	96-12-8	1,2-Dibromo-3-chloropropane
U044	67-66-3	Chloroform	U069	84-74-2	Dibutyl phthalate
U046	107-30-2	Chloromethyl methyl ether	U070	95-50-1	o-Dichlorobenzene
U047	91-58-7	beta-Chloronaphthalene	U071	541-73-1	m-Dichlorobenzene
U048	95-57-8	o-Chlorophenol	U072	106-46-7	p-Dichlorobenzene
U049	3165-93-3	4-Chloro-o-toluidine, hydrochloride	U073	91-94-1	3,3´-Dichlorobenzidine
U032	13765-19-0	Chromic acid H_2CrO_4, calcium salt	U074	764-41-0	1,4-Dichloro-2-butene (I,T)
U050	218-01-9	Chrysene	U075	75-71-8	Dichlorodifluoromethane
U051	NA	Creosote	U078	75-35-4	1,1-Dichloroethylene
U052	1319-77-3	Cresol (Cresylic acid)	U079	156-60-5	1,2-Dichloroethylene
U053	4170-30-3	Crotonaldehyde	U025	111-44-4	Dichloroethyl ether
U055	98-82-8	Cumeme (I)	U027	108-60-1	Dichloroisopropyl ether
U246	506-68-3	Cyanogen bromide (CN)Br	U024	111-91-1	Dichloromethoxy ethane
U197	106-51-4	2,5-Cyclohexadiene-1,4-dione	U081	120-83-2	2,4-Dichlorophenol
U056	110-82-7	Cyclohexane (I)	U082	87-65-0	2,6-Dichlorophenol
U129	58-89-9	Cyclohexane, 1,2,3,4,5,6-hexachloro-, (1alpha,2alpha,3beta,4alpha,5alpha,6beta)-	U084	542-75-6	1,3-Dichloropropene
			U085	1464-53-5	1,2:3,4-Diepoxybutane (I,T)
U057	108-94-1	Cyclohexanone (I)	U108	123-91-1	1,4-Diethyleneoxide
U130	77-47-4	1,3-Cyclopentadiene, 1,2,3,4,5,5-hexachloro-	U028	117-81-7	Diethylhexyl phthalate
			U395	5952-26-1	Diethylene glycol, dicarbamate
U058	50-18-0	Cyclophosphamide	U086	1615-80-1	N,N´-Diethylhydrazine
U240	94-75-7*	2,4-D, salts & esters	U087	3288-58-2	O,O-Diethyl S-methyl dithiophosphate
U059	20830-81-3	Daunomycin	U088	84-66-2	Diethyl phthalate
U060	72-54-8	DDD	U089	56-53-1	Diethylstilbesterol
U061	50-29-3	DDT	U090	94-58-6	Dihydrosafrole
U062	2303-16-4	Diallate	U091	119-90-4	3,3´-Dimethoxybenzidine
U063	53-70-3	Dibenz[a,h]anthracene	U092	124-40-3	Dimethylamine (I)

Table Appendix 1-1. (*continued*)

EPA Haz. Waste #	CAS* Number	Substance
U093	60-11-7	p-Dimethylaminoazobenzene
U094	57-97-6	7,12-Dimethylbenz[a]anthracene
U095	119-93-7	3,3´,-Dimethylbenzidine
U096	80-15-9	alpha,alpha-Dimethylbenzylhydroperoxide (R)
U097	79-44-7	Dimethylcarbamoyl chloride
U098	57-14-7	1,1-Dimethylhydrazine
U099	540-73-8	1,2-Dimethylhydrazine
U101	105-67-9	2,4-Dimethylphenol
U102	131-11-3	Dimethyl phthalate
U103	77-78-1	Dimethyl sulfate
U105	121-14-2	2,4-Dinitrotoluene
U106	606-20-2	2,6-Dinitrotoluene
U107	117-84-0	Di-n-octyl phthalate
U108	123-91-1	1,4-Dioxane
U109	122-66-7	1,2-Diphenylhydrazine
U110	142-84-7	Dipropylamine (I)
U111	621-64-7	Di-n-propylnitrosamine
U041	106-89-8	Epichlorohydrin
U001	75-07-0	Ethanal (I)
U404	121-44-8	Ethanamine, N,N-diethyl-
U174	55-18-5	Ethanamine, N-ethyl-N-nitroso-
U155	91-80-5	1,2-Ethanediamine, N,N-dimethyl-N´-2-pyridinyl-N´-(2-thienylmethyl)-
U067	106-93-4	Ethane, 1,2-dibromo-
U076	75-34-3	Ethane, 1,1-dichloro-
U077	107-06-2	Ethane, 1,2-dichloro-
U131	67-72-1	Ethane, hexachloro-

EPA Haz. Waste #	CAS* Number	Substance
U024	111-91-1	Ethane, 1,1´-[methylenebis(oxy)]bis[2-chloro-
U117	60-29-7	Ethane, 1,1´-oxybis- (I)
U025	111-44-4	Ethane, 1,1´-oxybis[2-chloro-
U184	76-01-7	Ethane, pentachloro-
U208	630-20-6	Ethane, 1,1,1,2-tetrachloro-
U209	79-34-5	Ethane, 1,1,2,2-tetrachloro-
U218	62-55-5	Ethanethioamide
U226	71-55-6	Ethane, 1,1,1-trichloro-
U227	79-00-5	Ethane, 1,1,2-trichloro-
U410	59669-26-0	Ethanimidothioic acid, N,N´-[thiobis[(methylimino)carbonyloxy]]bis-, dimethyl ester
U394	30558-43-1	Ethanimidothioic acid, 2-(dimethylamino)-N-hydroxy-2-oxo-, methyl ester
U359	110-80-5	Ethanol, 2-ethoxy-
U173	1116-54-7	Ethanol, 2,2´-(nitrosoimino)bis-
U395	5952-26-1	Ethanol, 2,2´-oxybis-, dicarbamate
U004	98-86-2	Ethanone, 1-phenyl-
U043	75-01-4	Ethene, chloro-
U042	110-75-8	Ethene, (2-chloroethoxy)-
U078	75-35-4	Ethene, 1,1-dichloro-
U079	156-60-5	Ethene, 1,2-dichloro-, (E)-
U210	127-18-4	Ethene, tetrachloro-
U228	79-01-6	Ethene, trichloro-
U112	141-78-6	Ethyl acetate (I)
U113	140-88-5	Ethyl acrylate (I)
U238	51-79-6	Ethyl carbamate (urethane)
U117	60-29-7	Ethyl ether (I)

Table Appendix 1-1. (*continued*)

EPA Haz. Waste #	CAS* Number	Substance
U114	111-54-6*	Ethylenebisdithiocarbamic acid, salts & esters
U067	106-93-4	Ethylene dibromide
U077	107-06-2	Ethylene dichloride
U359	110-80-5	Ethylene glycol monoethyl ether
U115	75-21-8	Ethylene oxide (I,T)
U116	96-45-7	Ethylenethiourea
U076	75-34-3	Ethylidene dichloride
U118	97-63-2	Ethyl methacrylate
U119	62-50-0	Ethyl methanesulfonate
U120	206-44-0	Fluoranthene
U122	50-00-0	Formaldehyde
U123	64-18-6	Formic acid (C,T)
U124	110-00-9	Furan (I)
U125	98-01-1	2-Furancarboxaldehyde (I)
U147	108-31-6	2,5-Furandione
U213	109-99-9	Furan, tetrahydro- (I)
U125	98-01-1	Furfural (I)
U124	110-00-9	Furfuran (I)
U206	18883-66-4	Glucopyranose, 2-deoxy-2-(3-methyl-3-nitrosoureido)-, D-
U206	18883-66-4	D-Glucose, 2-deoxy-2-[[(methylnitrosoamino)-carbonyl]amino]-
U126	765-34-4	Glycidylaldehyde
U163	70-25-7	Guanidine, N-methyl-N´-nitro-N-nitroso-
U127	118-74-1	Hexachlorobenzene
U128	87-68-3	Hexachlorobutadiene
U130	77-47-4	Hexachlorocyclopentadiene
U131	67-72-1	Hexachloroethane

EPA Haz. Waste #	CAS* Number	Substance
U132	70-30-4	Hexachlorophene
U243	1888-71-7	Hexachloropropene
U133	302-01-2	Hydrazine (R,T)
U086	1615-80-1	Hydrazine, 1,2-diethyl-
U098	57-14-7	Hydrazine, 1,1-dimethyl-
U099	540-73-8	Hydrazine, 1,2-dimethyl-
U109	122-66-7	Hydrazine, 1,2-diphenyl-
U134	7664-39-3	Hydrofluoric acid (C,T)
U134	7664-39-3	Hydrogen fluoride (C,T)
U135	7783-06-4	Hydrogen sulfide
U135	7783-06-4	Hydrogen sulfide H_2S
U096	80-15-9	Hydroperoxide, 1-methyl-1-phenylethyl- (R)
U116	96-45-7	2-Imidazolidinethione
U137	193-39-5	Indeno[1,2,3-cd]pyrene
U190	85-44-9	1,3-Isobenzofurandione
U140	78-83-1	Isobutyl alcohol (I,T)
U141	120-58-1	Isosafrole
U142	143-50-0	Kepone
U143	303-34-4	Lasiocarpine
U144	301-04-2	Lead acetate
U146	1335-32-6	Lead, bis(acetato-O)tetrahydroxytri-
U145	7446-27-7	Lead phosphate
U146	1335-32-6	Lead subacetate
U129	58-89-9	Lindane
U163	70-25-7	MNNG
U147	108-31-6	Maleic anhydride
U148	123-33-1	Maleic hydrazide
U149	109-77-3	Malononitrile

Table Appendix 1-1. (*continued*)

EPA Haz. Waste #	CAS* Number	Substance	EPA Haz. Waste #	CAS* Number	Substance
U150	148-82-3	Melphalan	U156	79-22-1	Methyl chlorocarbonate (I,T)
U151	7439-97-6	Mercury	U226	71-55-6	Methyl chloroform
U152	126-98-7	Methacrylonitrile (I,T)	U157	56-49-5	3-Methylcholanthrene
U092	124-40-3	Methanamine, N-methyl- (I)	U158	101-14-4	4,4′-Methylenebis(2-chloroaniline)
U029	74-83-9	Methane, bromo-	U068	74-95-3	Methylene bromide
U045	74-87-3	Methane, chloro- (I,T)	U080	75-09-2	Methylene chloride
U046	107-30-2	Methane, chloromethoxy-	U159	78-93-3	Methyl ethyl ketone (MEK) (I,T)
U068	74-95-3	Methane, dibromo-	U160	1338-23-4	Methyl ethyl ketone peroxide (R,T)
U080	75-09-2	Methane, dichloro-	U138	74-88-4	Methyl iodide
U075	75-71-8	Methane, dichlorodifluoro-	U161	108-10-1	Methyl isobutyl ketone (I)
U138	74-88-4	Methane, iodo-	U162	80-62-6	Methyl methacrylate (I,T)
U119	62-50-0	Methanesulfonic acid, ethyl ester	U161	108-10-1	4-Methyl-2-pentanone (I)
U211	56-23-5	Methane, tetrachloro-	U164	56-04-2	Methylthiouracil
U153	74-93-1	Methanethiol (I,T)	U010	50-07-7	Mitomycin C
U225	75-25-2	Methane, tribromo-	U059	20830-81-3	5,12-Naphthacenedione, 8-acetyl-10-[(3-amino-2,3,6-trideoxy)-alpha-L-lyxo-hexopyranosyl)oxy]-7,8,9,10-tetrahydro-6,8,11-trihydroxy-1-methoxy-, (8S-cis)-
U044	67-66-3	Methane, trichloro-			
U121	75-69-4	Methane, trichlorofluoro-			
U036	57-74-9	4,7-Methano-1H-indene, 1,2,4,5,6,7,8,8-octachloro-2,3,3a,4,7,7a-hexahydro-	U167	134-32-7	1-Naphthalenamine
U154	67-56-1	Methanol (I)	U168	91-59-8	2-Naphthalenamine
U155	91-80-5	Methapyrilene	U026	494-03-1	Naphthalenamine, N,N′-bis(2-chloroethyl)-
U142	143-50-0	1,3,4-Metheno-2H-cyclobuta[cd]pentalen-2-one, 1,1a,3,3a,4,5,5,5a,5b,6-decachlorooctahydro-	U165	91-20-3	Naphthalene
U247	72-43-5	Methoxychlor	U047	91-58-7	Naphthalene, 2-chloro-
U154	67-56-1	Methyl alcohol (I)	U166	130-15-4	1,4-Naphthalenedione
U029	74-83-9	Methyl bromide	U236	72-57-1	2,7-Naphthalenedisulfonic acid, 3,3′-[(3,3′-dimethyl[1,1′-biphenyl]-4,4′-diyl)-bis(azo)bis[5-amino-4-hydroxy]-, tetrasodium salt
U186	504-60-9	1-Methylbutadiene (I)			
U045	74-87-3	Methyl chloride (I,T)	U279	63-25-2	1-Naphthalenol, methylcarbamate
			U166	130-15-4	1,4-Naphthoquinone

Table Appendix 1-1. (*continued*)

EPA Haz. Waste #	CAS* Number	Substance	EPA Haz. Waste #	CAS* Number	Substance
U167	134-32-7	alpha-Naphthylamine	U187	62-44-2	Phenacetin
U168	91-59-8	beta-Naphthylamine	U188	108-95-2	Phenol
U217	10102-45-1	Nitric acid, thallium(1+) salt	U048	95-57-8	Phenol, 2-chloro-
U169	98-95-3	Nitrobenzene (I,T)	U039	59-50-7	Phenol, 4-chloro-3-methyl-
U170	100-02-7	p-Nitrophenol	U081	120-83-2	Phenol, 2,4-dichloro-
U171	79-46-9	2-Nitropropane (I,T)	U082	87-65-0	Phenol, 2,6-dichloro-
U172	924-16-3	N-Nitrosodi-n-butylamine	U089	56-53-1	Phenol, 4,4´-(1,2-diethyl-1,2-ethenediyl)bis-, (E)-
U173	1116-54-7	N-Nitrosodiethanolamine	U101	105-67-9	Phenol, 2,4-dimethyl-
U174	55-18-5	N-Nitrosodiethylamine	U052	1319-77-3	Phenol, methyl-
U176	759-73-9	N-Nitroso-N-ethylurea	U132	70-30-4	Phenol, 2,2´-methylenebis[3,4,6-trichloro-
U177	684-93-5	N-Nitroso-N-methylurea	U411	114-26-1	Phenol, 2-(1-methylethoxy)-, methylcarbamate
U178	615-53-2	N-Nitroso-N-methylurethane	U170	100-02-7	Phenol, 4-nitro-
U179	100-75-4	N-Nitrosopiperidine	See F027	87-86-5	Phenol, pentachloro-
U180	930-55-2	N-Nitrosopyrrolidine	See F027	58-90-2	Phenol, 2,3,4,6-tetrachloro-
U181	99-55-8	5-Nitro-o-toluidine	See F027	95-95-4	Phenol, 2,4,5-trichloro-
U193	1120-71-4	1,2,-Oxathiolane, 2,2-dioxide	See F027	88-06-2	Phenol, 2,4,6-trichloro-
U058	50-18-0	2H-1,3,2-Oxazaphosphorin-2-amine, N,N-bis(2-chloroethyl)tetrahydro-, 2-oxide	U150	148-82-3	L-Phenylalanine, 4-[bis(2-chloroethyl)amino]-
U115	75-21-8	Oxirane (I,T)	U145	7446-27-7	Phosphoric acid, lead(2+) salt (2:3)
U126	765-34-4	Oxiranecarboxyaldehyde	U087	3288-58-2	Phosphorodithioic acid, O,O-diethyl S-methyl ester
U041	106-89-8	Oxirane, (chloromethyl)-	U189	1314-80-3	Phosphorus sulfide (R)
U182	123-63-7	Paraldehyde	U190	85-44-9	Phthalic anhydride
U183	603-93-5	Pentachlorobenzene	U191	109-06-8	2-Picoline
U184	76-01-7	Pentachloroethane	U179	100-75-4	Piperidine, 1-nitroso-
U185	82-68-8	Pentachloronitrobenzene (PCNB)	U192	23950-58-5	Pronamide
See F027	87-86-5	Pentachlorophenol	U194	107-10-8	1-Propanamine (I,T)
U161	108-10-1	Pentanol, 4-methyl-			
U186	504-60-9	1,3-Pentadiene (I)			

Table Appendix 1-1. (*continued*)

EPA Haz. Waste #	CAS* Number	Substance	EPA Haz. Waste #	CAS* Number	Substance
U111	621-64-7	1-Propanamine, N-nitroso-N-propyl-	U148	123-33-1	3,6-Pyridazinedione, 1,2-dihydro-
U110	142-84-7	1-Propanamine, N-propyl- (I)	U196	110-86-1	Pyridine
U066	96-12-8	Propane, 1,2-dibromo-3-chloro-	U191	109-06-8	Pyridine, 2-methyl-
U083	78-87-5	Propane, 1,2-dichloro-	U237	66-75-1	2,4-(1H,3H)-Pyrimidinedione, 5-[bis(2-chloroethyl)amino]-
U149	109-77-3	Propanedinitrile	U164	56-04-2	4(1H)-Pyrimidinone, 2,3-dihydro-6-methyl-2-thioxo-
U171	79-46-9	Propane, 2-nitro- (I,T)			
U027	108-60-1	Propane, 2,2′-oxybis[2-chloro-	U180	930-55-2	Pyrrolidine, 1-nitroso-
U193	1120-71-4	1,3-Propane sultone	U200	50-55-5	Reserpine
See F027	93-72-1	Propanoic acid, 2-(2,4,5-trichlorophenoxy)-	U201	108-46-3	Resorcinol
			U203	94-59-7	Safrole
U235	126-72-7	1-Propanol, 2,3-dibromo-, phosphate (3:1)	U204	7783-00-8	Selenious acid
U140	78-83-1	1-Propanol, 2-methyl- (I,T)	U204	7783-00-8	Selenium dioxide
U002	67-64-1	2-Propanone (I)	U205	7488-56-4	Selenium sulfide
U007	79-06-1	2-Propenamide	U205	7488-56-4	Selenium sulfide SeS₂(R,T)
U084	542-75-6	Propene, 1,3-dichloro-	U015	115-02-6	L-Serine, diazoacetate (ester)
U243	1888-71-7	1-Propene, 1,1,2,3,3,3-hexachloro-	See F027	93-72-1	Silvex (2,4,5-TP)
U009	107-13-1	2-Propenenitrile	U206	18883-66-4	Streptozotocin
U152	126-98-7	2-Propenenitrile, 2-methyl- (I,T)	U103	77-78-1	Sulfuric acid, dimethyl ester
U008	79-10-7	2-Propenoic acid (I)	U189	1314-80-3	Sulfur phosphide (R)
U113	140-88-5	2-Propenoic acid, ethyl ester (I)	See F027	93-76-5	2,4,5-T
U118	97-63-2	2-Propenoic acid, 2-methyl-, ethyl ester	U207	95-94-3	1,2,4,5-Tetrachlorobenzene
U162	80-62-6	2-Propenoic acid, 2-methyl-, methyl ester (I,T)	U208	630-20-6	1,1,1,2-Tetrachloroethane
			U209	79-34-5	1,1,2,2-Tetrachloroethane
U373	122-42-9	Propham	U210	127-18-4	Tetrachloroethylene
U411	114-26-1	Propoxur	See F027	58-90-2	2,3,4,6-Tetrachlorophenol
U387	52888-80-9	Prosulfocarb	U213	109-99-9	Tetrahydrofuran (I)
U194	107-10-8	n-Propylamine (I,T)	U214	563-68-8	Thallium(I) acetate
U083	78-87-5	Propylene dichloride	U215	6533-73-9	Thallium(I) carbonate

Table Appendix 1-1. (*continued*)

EPA Haz. Waste #	CAS* Number	Substance
U216	7791-12-0	Thallium(I) chloride
U216	7791-12-0	Thallium chloride TlCl
U217	10102-45-1	Thallium(I) nitrate
U218	62-55-5	Thioacetamide
U410	59669-26-0	Thiodicarb
U153	74-93-1	Thiomethanol (I,T)
U244	137-26-8	Thioperoxydicarbonic diamide [(H$_2$N) C(S)]$_2$S$_2$, tetramethyl-
U409	23564-05-8	Thiophanate-methyl
U219	62-56-6	Thiourea
U244	137-26-8	Thiram
U220	108-88-3	Toluene
U221	25376-45-8	Toluenediamine
U223	26471-62-5	Toluene diisocyanate (R,T)
U328	95-53-4	o-Toluidine
U353	106-49-0	p-Toluidine
U222	636-21-5	o-Toluidine hydrochloride
U389	2303-17-5	Triallate
U011	61-82-5	1H-1,2,4-Triazol-3-amine
U226	71-55-6	1,1,1-Trichloroethane
U227	79-00-5	1,1,2-Trichloroethane
U228	79-01-6	Trichloroethylene
U121	75-69-4	Trichloromonofluoromethane
See F027	95-95-4	2,4,5-Trichlorophenol
See F027	88-06-2	2,4,6-Trichlorophenol
U404	121-44-8	Triethylamine
U234	99-35-4	1,3,5-Trinitrobenzene (R,T)
U182	123-63-7	1,3,5-Trioxane, 2,4,6-trimethyl-
U235	126-72-7	Tris(2,3-dibromopropyl) phosphate

EPA Haz. Waste #	CAS* Number	Substance
U236	72-57-1	Trypan blue
U237	66-75-1	Uracil mustard
U176	759-73-9	Urea, N-ethyl-N-nitroso-
U177	684-93-5	Urea, N-methyl-N-nitroso-
U043	75-01-4	Vinyl chloride
U248	81-81-2*	Warfarin, & salts, when present at concentrations of 0.3% or less
U239	1330-20-7	Xylene (I)
U200	50-55-5	Yohimban-16-carboxylic acid, 11,17-dimethoxy-18-[(3,4,5-trimethoxy benzoyl)oxy]-, methyl ester, (3beta, 16beta,17alpha,18beta,20alpha)-
U249	1314-84-7	Zinc phosphide Zn$_3$P$_2$, when present at concentrations of 10% or less

Notes

*CAS Number: Given for parent compound only.

NA: Not applicable.

P-listed Waste: The primary hazardous properties of these materials is indicated by the letters T (toxicity) and R (reactivity). Absence of a letter indicates that the compound only is listed for acute toxicity.

U-listed Waste: The primary hazardous properties of these materials is indicated by the letters T (toxicity), R (reactivity), I (ignitability), and C (corrosivity). Absence of a letter indicates that the compound is only listed for toxicity.

Source: http://ecfr.gpoaccess.gov/cgi/t/text/text-idx?c=ecfr&sid=f3fc1aeedcb0cc87bf4367f0a9cc03d0&rgn=div5&view=text&node=40:26.0.1.1.2&idno=40#40:26.0.1.1.2.4.1.4

Table Appendix 1-2. Discarded Commercial Chemical Products (P & U Lists): Alphanumeric by EPA Hazardous Waste #

EPA Haz. Waste #	CAS* Number	Substance	EPA Haz. Waste #	CAS* Number	Substance
— Acutely Hazardous §261.33(e) —			P014	108-98-5	Benzenethiol
P001	81-81-2*	2H-1-Benzopyran-2-one, 4-hydroxy-3-(3-oxo-1-phenylbutyl)-, & salts, when present at concentrations greater than 0.3%	P014	108-98-5	Thiophenol
			P015	7440-41-7	Beryllium powder
			P016	542-88-1	Dichloromethyl ether
P001	81-81-2*	Warfarin, & salts, when present at concentrations greater than 0.3%	P016	542-88-1	Methane, oxybis[chloro-
P002	591-08-2	1-Acetyl-2-thiourea	P017	598-31-2	2-Propanone, 1-bromo-
P002	591-08-2	Acetamide, N-(aminothioxomethyl)-	P017	598-31-2	Bromoacetone
P003	107-02-8	2-Propenal	P018	357-57-3	Brucine
P003	107-02-8	Acrolein	P018	357-57-3	Strychnidin-10-one, 2,3-dimethoxy-
P004	309-00-2	1,4,5,8-Dimethanonaphthalene, 1,2,3,4,10,10-hexachloro-1,4,4a,5,8,8a-hexahydro-, (1alpha, 4alpha,4abeta,5alpha,8alpha,8abeta)-	P020	88-85-7	Dinoseb
			P020	88-85-7	Phenol, 2-(1-methylpropyl)-4,6-dinitro-
P004	309-00-2	Aldrin	P021	592-01-8	Calcium cyanide
P005	107-18-6	2-Propen-1-ol	P021	592-01-8	Calcium cyanide Ca(CN)₂
P005	107-18-6	Allyl alcohol	P022	75-15-0	Carbon disulfide
P006	20859-73-8	Aluminum phosphide (R,T)	P023	107-20-0	Acetaldehyde, chloro-
P007	2763-96-4	3(2H)-Isoxazolone, 5-(aminomethyl)-	P023	107-20-0	Chloroacetaldehyde
P007	2763-96-4	5-(Aminomethyl)-3-isoxazolol	P024	106-47-8	Benzenamine, 4-chloro-
P008	504-24-5	4-Aminopyridine	P024	106-47-8	p-Chloroaniline
P008	504-24-5	4-Pyridinamine	P026	5344-82-1	1-(o-Chlorophenyl)thiourea
P009	131-74-8	Ammonium picrate (R)	P026	5344-82-1	Thiourea, (2-chlorophenyl)-
P009	131-74-8	Phenol, 2,4,6-trinitro-, ammonium salt (R)	P027	542-76-7	3-Chloropropionitrile
			P027	542-76-7	Propanenitrile, 3-chloro-
P010	7778-39-4	Arsenic acid H₃AsO₄	P028	100-44-7	Benzene, (chloromethyl)-
P011	1303-28-2	Arsenic oxide As₂O₅	P028	100-44-7	Benzyl chloride
P011	1303-28-2	Arsenic pentoxide	P029	544-92-3	Copper cyanide
P012	1327-53-3	Arsenic oxide As₂O₃	P029	544-92-3	Copper cyanide Cu(CN)
P012	1327-53-3	Arsenic trioxide	P030	NA	Cyanides (soluble cyanide salts), not otherwise specified
P013	542-62-1	Barium cyanide			

Table Appendix 1-2. (*continued*)

EPA Haz. Waste #	CAS* Number	Substance
P031	460-19-5	Cyanogen
P031	460-19-5	Ethanedinitrile
P033	506-77-4	Cyanogen chloride
P033	506-77-4	Cyanogen chloride (CN)Cl
P034	131-89-5	2-Cyclohexyl-4,6-dinitrophenol
P034	131-89-5	Phenol, 2-cyclohexyl-4,6-dinitro-
P036	696-28-6	Arsonous dichloride, phenyl-
P036	696-28-6	Dichlorophenylarsine
P037	60-57-1	2,7:3,6-Dimethanonaphth[2,3-b] oxirene, 3,4,5,6,9,9-hexachloro-1a,2, 2a,3,6,6a,7,7a-octahydro-, (1aalpha, 2beta,2aalpha,3beta,6beta,6aalpha, 7beta,7aalpha)-
P037	60-57-1	Dieldrin
P038	692-42-2	Arsine, diethyl-
P038	692-42-2	Diethylarsine
P039	298-04-4	Disulfoton
P039	298-04-4	Phosphorodithioic acid, O,O-diethyl S-[2-(ethylthio)ethyl] ester
P040	297-97-2	O,O-Diethyl O-pyrazinyl phosphorothioate
P040	297-97-2	Phosphorothioic acid, O,O-diethyl O-pyrazinyl ester
P041	311-45-5	Diethyl-p-nitrophenyl phosphate
P041	311-45-5	Phosphoric acid, diethyl 4-nitrophenyl ester
P042	51-43-4	1,2-Benzenediol, 4-[1-hydroxy-2-(methylamino)ethyl]-, (R)-
P042	51-43-4	Epinephrine
P043	55-91-4	Diisopropylfluorophosphate (DFP)
P043	55-91-4	Phosphorofluoridic acid, bis(1-methylethyl) ester
P044	60-51-5	Dimethoate

EPA Haz. Waste #	CAS* Number	Substance
P044	60-51-5	Phosphorodithioic acid, O,O-dimethyl S-[2-(methylamino)-2-oxoethyl] ester
P045	39196-18-4	2-Butanone, 3,3-dimethyl-1-(methylthio)-, O-[(methylamino) carbonyl] oxime
P045	39196-18-4	Thiofanox
P046	122-09-8	alpha,alpha-Dimethylphenethylamine
P046	122-09-8	Benzeneethanamine, alpha,alpha-dimethyl-
P047	534-52-1*	4,6-Dinitro-o-cresol, & salts
P047	534-52-1*	Phenol, 2-methyl-4,6-dinitro-, & salts
P048	51-28-5	2,4-Dinitrophenol
P048	51-28-5	Phenol, 2,4-dinitro-
P049	541-53-7	Dithiobiuret
P049	541-53-7	Thioimidodicarbonic diamide [(H$_2$N)C(S)]$_2$NH
P050	115-29-7	6,9-Methano-2,4,3-benzodioxathiepin, 6,7,8,9,10,10-hexachloro-1,5,5a,6,9,9a-hexahydro-, 3-oxide
P050	115-29-7	Endosulfan
P051	72-20-8	Endrin
P051	72-20-8	Endrin & metabolites
P051	72-20-8*	2,7:3,6-Dimethanonaphth[2,3-b] oxirene, 3,4,5,6,9,9-hexachloro-1a,2, 2a,3,6,6a,7,7a-octahydro-, (1aalpha, 2beta,2abeta,3alpha,6alpha,6abeta, 7beta,7aalpha)-, & metabolites
P054	151-56-4	Aziridine
P054	151-56-4	Ethyleneimine
P056	7782-41-4	Fluorine
P057	640-19-7	Acetamide, 2-fluoro-
P057	640-19-7	Fluoroacetamide
P058	62-74-8	Acetic acid, fluoro-, sodium salt

Table Appendix 1-2. (*continued*)

EPA Haz. Waste #	CAS* Number	Substance	EPA Haz. Waste #	CAS* Number	Substance
P058	62-74-8	Fluoroacetic acid, sodium salt	P071	298-00-0	Phosphorothioic acid, O,O,-dimethyl O-(4-nitrophenyl) ester
P059	76-44-8	4,7-Methano-1H-indene, 1,4,5,6,7,8, 8-heptachloro-3a,4,7,7a-tetrahydro-	P072	86-88-4	alpha-Naphthylthiourea
P059	76-44-8	Heptachlor	P072	86-88-4	Thiourea, 1-naphthalenyl-
P060	465-73-6	1,4,5,8-Dimethanonaphthalene, 1,2,3,4,10,10-hexachloro-1,4,4a,5,8,8a-hexahydro-, (1alpha, 4alpha,4abeta,5beta,8beta,8abeta)-	P073	13463-39-3	Nickel carbonyl
			P073	13463-39-3	Nickel carbonyl Ni(CO)$_4$, (T-4)-
			P074	557-19-7	Nickel cyanide
P060	465-73-6	Isodrin	P074	557-19-7	Nickel cyanide Ni(CN)$_2$
P062	757-58-4	Hexaethyl tetraphosphate	P075	54-11-5*	Nicotine, & salts
P062	757-58-4	Tetraphosphoric acid, hexaethyl ester	P075	54-11-5*	Pyridine, 3-(1-methyl-2-pyrrolidinyl)-, (S)-, & salts
P063	74-90-8	Hydrocyanic acid			
P063	74-90-8	Hydrogen cyanide	P076	10102-43-9	Nitric oxide
P064	624-83-9	Methane, isocyanato-	P076	10102-43-9	Nitrogen oxide NO
P064	624-83-9	Methyl isocyanate	P077	100-01-6	Benzenamine, 4-nitro-
P065	628-86-4	Fulminic acid, mercury(2+) salt (R,T)	P077	100-01-6	p-Nitroaniline
P065	628-86-4	Mercury fulminate (R,T)	P078	10102-44-0	Nitrogen dioxide
P066	16752-77-5	Ethanimidothioic acid, N-[[(methyl amino)carbonyl]oxy]-, methyl ester	P078	10102-44-0	Nitrogen oxide NO$_2$
			P081	55-63-0	1,2,3-Propanetriol, trinitrate (R)
P066	16752-77-5	Methomyl	P081	55-63-0	Nitroglycerine (R)
P067	75-55-8	1,2-Propylenimine	P082	62-75-9	Methanamine, N-methyl-N-nitroso-
P067	75-55-8	Aziridine, 2-methyl-	P082	62-75-9	N-Nitrosodimethylamine
P068	60-34-4	Hydrazine, methyl-	P084	4549-40-0	N-Nitrosomethylvinylamine
P068	60-34-4	Methyl hydrazine	P084	4549-40-0	Vinylamine, N-methyl-N-nitroso-
P069	75-86-5	2-Methyllactonitrile	P085	152-16-9	Diphosphoramide, octamethyl-
P069	75-86-5	Propanenitrile, 2-hydroxy-2-methyl-	P085	152-16-9	Octamethylpyrophosphoramide
P070	116-06-3	Aldicarb	P087	20816-12-0	Osmium oxide OsO$_4$, (T-4)-
P070	116-06-3	Propanal, 2-methyl-2-(methylthio)-, O-[(methylamino)carbonyl]oxime	P087	20816-12-0	Osmium tetroxide
P071	298-00-0	Methyl parathion	P088	145-73-3	7-Oxabicyclo[2.2.1]heptane-2,3-dicarboxylic acid

Table Appendix 1-2. (*continued*)

EPA Haz. Waste #	CAS* Number	Substance
P088	145-73-3	Endothall
P089	56-38-2	Parathion
P089	56-38-2	Phosphorothioic acid, O,O-diethyl O-(4-nitrophenyl) ester
P092	62-38-4	Mercury, (acetato-O)phenyl-
P092	62-38-4	Phenylmercury acetate
P093	103-85-5	Phenylthiourea
P093	103-85-5	Thiourea, phenyl-
P094	298-02-2	Phorate
P094	298-02-2	Phosphorodithioic acid, O,O-diethyl S-[(ethylthio)methyl] ester
P095	75-44-5	Carbonic dichloride
P095	75-44-5	Phosgene
P096	7803-51-2	Hydrogen phosphide
P096	7803-51-2	Phosphine
P097	52-85-7	Famphur
P097	52-85-7	Phosphorothioic acid, O-[4-[(dimethylamino)sulfonyl]phenyl] O,O-dimethyl ester
P098	151-50-8	Potassium cyanide
P098	151-50-8	Potassium cyanide K(CN)
P099	506-61-6	Argentate(1-), bis(cyano-C)-, potassium
P099	506-61-6	Potassium silver cyanide
P101	107-12-0	Ethyl cyanide
P101	107-12-0	Propanenitrile
P102	107-19-7	2-Propyn-1-ol
P102	107-19-7	Propargyl alcohol
P103	630-10-4	Selenourea
P104	506-64-9	Silver cyanide

EPA Haz. Waste #	CAS* Number	Substance
P104	506-64-9	Silver cyanide Ag(CN)
P105	26628-22-8	Sodium azide
P106	143-33-9	Sodium cyanide
P106	143-33-9	Sodium cyanide Na(CN)
P108	57-24-9*	Strychnidin-10-one, & salts
P108	57-24-9*	Strychnine, & salts
P109	3689-24-5	Tetraethyldithiopyrophosphate
P109	3689-24-5	Thiodiphosphoric acid, tetraethyl ester
P110	78-00-2	Plumbane, tetraethyl-
P110	78-00-2	Tetraethyl lead
P111	107-49-3	Diphosphoric acid, tetraethyl ester
P111	107-49-3	Tetraethyl pyrophosphate
P112	509-14-8	Methane, tetranitro- (R)
P112	509-14-8	Tetranitromethane (R)
P113	1314-32-5	Thallic oxide
P113	1314-32-5	Thallium oxide Tl_2O_3
P114	12039-52-0	Selenious acid, dithallium(1+) salt
P114	12039-52-0	Thallium(I) selenite
P115	7446-18-6	Sulfuric acid, dithallium(1+) salt
P115	7446-18-6	Thallium(I) sulfate
P116	79-19-6	Hydrazinecarbothioamide
P116	79-19-6	Thiosemicarbazide
P118	75-70-7	Methanethiol, trichloro-
P118	75-70-7	Trichloromethanethiol
P119	7803-55-6	Ammonium vanadate
P119	7803-55-6	Vanadic acid, ammonium salt
P120	1314-62-1	Vanadium oxide V_2O_5
P120	1314-62-1	Vanadium pentoxide

Table Appendix 1-2. (*continued*)

EPA Haz. Waste #	CAS* Number	Substance	EPA Haz. Waste #	CAS* Number	Substance
P121	557-21-1	Zinc cyanide	P194	23135-22-0	Ethanimidothioic acid, 2-(dimethylamino)-N-[[(methylamino) carbonyl]oxy]-2-oxo-, methyl ester
P121	557-21-1	Zinc cyanide Zn(CN)$_2$			
P122	1314-84-7	Zinc phosphide Zn$_3$P$_2$, when present at concentrations greater than 10% (R,T)	P194	23135-22-0	Oxamyl
			P196	15339-36-3	Manganese dimethyldithiocarbamate
P123	8001-35-2	Toxaphene	P196	15339-36-3	Manganese, bis(dimethylcarbamodithioato-S,S′)-,
P127	1563-66-2	7-Benzofuranol, 2,3-dihydro-2,2-dimethyl-, methylcarbamate			
			P197	17702-57-7	Formparanate
P127	1563-66-2	Carbofuran	P197	17702-57-7	Methanimidamide, N,N-dimethyl-N′-[2-methyl-4-[[(methylamino)carbonyl] oxy]phenyl]-
P128	315-18-4	Phenol, 4-(dimethylamino)-3,5-dimethyl-, methylcarbamate (ester)			
P128	315-8-4	Mexacarbate	P198	23422-53-9	Formetanate hydrochloride
P185	26419-73-8	1,3-Dithiolane-2-carboxaldehyde, 2,4-dimethyl-, O- [(methylamino)-carbonyl]oxime	P198	23422-53-9	Methanimidamide, N,N-dimethyl-N′-[3-[[(methylamino)-carbonyl]oxy] phenyl]-, monohydrochloride
P185	26419-73-8	Tirpate	P199	2032-65-7	Methiocarb
P188	57-64-7	Benzoic acid, 2-hydroxy-, compd. with (3aS-cis)-1,2,3,3a,8,8a-hexahydro-1,3a, 8-trimethylpyrrolo[2,3-b]indol-5-yl methylcarbamate ester (1:1)	P199	2032-65-7	Phenol, (3,5-dimethyl-4-(methylthio)-, methylcarbamate
			P201	2631-37-0	Phenol, 3-methyl-5-(1-methylethyl)-, methyl carbamate
P188	57-64-7	Physostigmine salicylate	P201	2631-37-0	Promecarb
P189	55285-14-8	Carbamic acid, [(dibutylamino)- thio]methyl-, 2,3-dihydro-2,2-dimethyl-7-benzofuranyl ester	P202	64-00-6	3-Isopropylphenyl N-methylcarbamate
			P202	64-00-6	m-Cumenyl methylcarbamate
P189	55285-14-8	Carbosulfan	P202	64-00-6	Phenol, 3-(1-methylethyl)-, methyl carbamate
P190	1129-41-5	Carbamic acid, methyl-, 3-methylphenyl ester	P203	1646-88-4	Aldicarb sulfone
P190	1129-41-5	Metolcarb	P203	1646-88-4	Propanal, 2-methyl-2-(methyl-sulfonyl)-, O-[(methylamino)carbonyl] oxime
P191	644-64-4	Carbamic acid, dimethyl-, 1-[(dimethyl-amino)carbonyl]-5-methyl-1H-pyrazol-3-yl ester			
			P204	57-47-6	Physostigmine
P191	644-64-4	Dimetilan	P204	57-47-6	Pyrrolo[2,3-b]indol-5-ol, 1,2,3,3a,8,8a-hexahydro-1,3a,8-trimethyl-, methylcarbamate (ester), (3aS-cis)-
P192	119-38-0	Carbamic acid, dimethyl-, 3-methyl-1-(1-methylethyl)-1H-pyrazol-5-yl ester			
P192	119-38-0	Isolan	P205	137-30-4	Zinc, bis(dimethylcarbamodithioato-S,S′)-,

Table Appendix 1-2. (*continued*)

EPA Haz. Waste #	CAS* Number	Substance
P205	137-30-4	Ziram
	— **Hazardous §261.33(f)** —	
U001	75-07-0	Acetaldehyde (I)
U001	75-07-0	Ethanal (I)
U002	67-64-1	2-Propanone (I)
U002	67-64-1	Acetone (I)
U003	75-05-8	Acetonitrile (I,T)
U004	98-86-2	Acetophenone
U004	98-86-2	Ethanone, 1-phenyl-
U005	53-96-3	2-Acetylaminofluorene
U005	53-96-3	Acetamide, N-9H-fluoren-2-yl-
U006	75-36-5	Acetyl chloride (C,R,T)
U007	79-06-1	2-Propenamide
U007	79-06-1	Acrylamide
U008	79-10-7	2-Propenoic acid (I)
U008	79-10-7	Acrylic acid (I)
U009	107-13-1	2-Propenenitrile
U009	107-13-1	Acrylonitrile
U010	50-07-7	Azirino[2′,3′:3,4]pyrrolo[1,2-a]indole-4,7-dione, 6-amino-8-[[(amino carbonyl)oxy]methyl]-1,1a,2,8,8a, 8b-hexahydro-8a-methoxy-5-methyl-, [1aS-(1aalpha,8beta,8aalpha,8balpha)]-
U010	50-07-7	Mitomycin C
U011	61-82-5	1H-1,2,4-Triazol-3-amine
U011	61-82-5	Amitrole
U012	62-53-3	Aniline (I,T)
U012	62-53-3	Benzenamine (I,T)
U014	492-80-8	Auramine

EPA Haz. Waste #	CAS* Number	Substance
U014	492-80-8	Benzenamine, 4,4′,-carbonimidoylbis[N,N-dimethyl-
U015	115-02-6	Azaserine
U015	115-02-6	L-Serine, diazoacetate (ester)
U016	225-51-4	Benz[c]acridine
U017	98-87-3	Benzal chloride
U017	98-87-3	Benzene, (dichloromethyl)-
U018	56-55-3	Benz[a]anthracene
U019	71-43-2	Benzene (I,T)
U020	98-09-9	Benzenesulfonic acid chloride (C,R)
U020	98-09-9	Benzenesulfonyl chloride (C,R)
U021	92-87-5	[1,1′-Biphenyl]-4,4′-diamine
U021	92-87-5	Benzidine
U022	50-32-8	Benzo[a]pyrene
U023	98-07-7	Benzene, (trichloromethyl)-
U023	98-07-7	Benzotrichloride (C,R,T)
U024	111-91-1	Dichloromethoxy ethane
U024	111-91-1	Ethane, 1,1′-[methylenebis(oxy)] bis[2-chloro-
U025	111-44-4	Dichloroethyl ether
U025	111-44-4	Ethane, 1,1′-oxybis[2-chloro-
U026	494-03-1	Chlornaphazin
U026	494-03-1	Naphthalenamine, N,N′-bis(2-chloroethyl)-
U027	108-60-1	Dichloroisopropyl ether
U027	108-60-1	Propane, 2,2′-oxybis[2-chloro-
U028	117-81-7	1,2-Benzenedicarboxylic acid, bis(2-ethylhexyl) ester
U028	117-81-7	Diethylhexyl phthalate
U029	74-83-9	Methane, bromo-

Table Appendix 1-2. (*continued*)

EPA Haz. Waste #	CAS* Number	Substance	EPA Haz. Waste #	CAS* Number	Substance
U029	74-83-9	Methyl bromide	U043	75-01-4	Vinyl chloride
U030	101-55-3	4-Bromophenyl phenyl ether	U044	67-66-3	Chloroform
U030	101-55-3	Benzene, 1-bromo-4-phenoxy-	U044	67-66-3	Methane, trichloro-
U031	71-36-3	1-Butanol (I)	U045	74-87-3	Methane, chloro- (I,T)
U031	71-36-3	n-Butyl alcohol (I)	U045	74-87-3	Methyl chloride (I,T)
U032	13765-19-0	Calcium chromate	U046	107-30-2	Chloromethyl methyl ether
U032	13765-19-0	Chromic acid H_2CrO_4, calcium salt	U046	107-30-2	Methane, chloromethoxy-
U033	353-50-4	Carbon oxyfluoride (R,T)	U047	91-58-7	beta-Chloronaphthalene
U033	353-50-4	Carbonic difluoride	U047	91-58-7	Naphthalene, 2-chloro-
U034	75-87-6	Acetaldehyde, trichloro-	U048	95-57-8	o-Chlorophenol
U034	75-87-6	Chloral	U048	95-57-8	Phenol, 2-chloro-
U035	305-03-3	Benzenebutanoic acid, 4-[bis(2-chloroethyl)amino]-	U049	3165-93-3	4-Chloro-o-toluidine, hydrochloride
U035	305-03-3	Chlorambucil	U049	3165-93-3	Benzenamine, 4-chloro-2-methyl-, hydrochloride
U036	57-74-9	4,7-Methano-1H-indene, 1,2,4,5,6,7,8,8-octachloro-2,3,3a,4,7,7a-hexahydro-	U050	218-01-9	Chrysene
			U051	NA	Creosote
U036	57-74-9	Chlordane, alpha & gamma isomers	U052	1319-77-3	Cresol (Cresylic acid)
U037	108-90-7	Benzene, chloro-	U052	1319-77-3	Phenol, methyl-
U037	108-90-7	Chlorobenzene	U053	4170-30-3	2-Butenal
U038	510-15-6	Benzeneacetic acid, 4-chloro-alpha-(4-chlorophenyl)-alpha-hydroxy-, ethyl ester	U053	4170-30-3	Crotonaldehyde
			U055	98-82-8	Benzene, (1-methylethyl)- (I)
U038	510-15-6	Chlorobenzilate	U055	98-82-8	Cumeme (I)
U039	59-50-7	p-Chloro-m-cresol	U056	110-82-7	Benzene, hexahydro- (I)
U039	59-50-7	Phenol, 4-chloro-3-methyl-	U056	110-82-7	Cyclohexane (I)
U041	106-89-8	Epichlorohydrin	U057	108-94-1	Cyclohexanone (I)
U041	106-89-8	Oxirane, (chloromethyl)-	U058	50-18-0	2H-1,3,2-Oxazaphosphorin-2-amine, N,N-bis(2-chloroethyl)tetrahydro-, 2-oxide
U042	110-75-8	2-Chloroethyl vinyl ether			
U042	110-75-8	Ethene, (2-chloroethoxy)-	U058	50-18-0	Cyclophosphamide
U043	75-01-4	Ethene, chloro-			

Table Appendix 1-2. (*continued*)

EPA Haz. Waste #	CAS* Number	Substance
U059	20830-81-3	5,12-Naphthacenedione, 8-acetyl-10-[(3-amino-2,3,6-trideoxy)-alpha-L-lyxo-hexopyranosyl)oxy]-7,8,9,10-tetrahydro-6,8,11-trihydroxy-1-methoxy-, (8S-cis)-
U059	20830-81-3	Daunomycin
U060	72-54-8	Benzene, 1,1'-(2,2-dichloroethylidene) bis[4-chloro-
U060	72-54-8	DDD
U061	50-29-3	Benzene, 1,1'-(2,2,2-trichloro ethylidene)bis[4-chloro-
U061	50-29-3	DDT
U062	2303-16-4	Carbamodithioic acid, bis(1-methylethyl)-, S-(2,3-dichloro-2-propenyl) ester
U062	2303-16-4	Diallate
U063	53-70-3	Dibenz[a,h]anthracene
U064	189-55-9	Benzo[rst]pentaphene
U064	189-55-9	Dibenzo[a,i]pyrene
U066	96-12-8	1,2-Dibromo-3-chloropropane
U066	96-12-8	Propane, 1,2-dibromo-3-chloro-
U067	106-93-4	Ethane, 1,2-dibromo-
U067	106-93-4	Ethylene dibromide
U068	74-95-3	Methane, dibromo-
U068	74-95-3	Methylene bromide
U069	84-74-2	1,2-Benzenedicarboxylic acid, dibutyl ester
U069	84-74-2	Dibutyl phthalate
U070	95-50-1	Benzene, 1,2-dichloro-
U070	95-50-1	o-Dichlorobenzene
U071	541-73-1	Benzene, 1,3-dichloro-
U071	541-73-1	m-Dichlorobenzene

EPA Haz. Waste #	CAS* Number	Substance
U072	106-46-7	Benzene, 1,4-dichloro-
U072	106-46-7	p-Dichlorobenzene
U073	91-94-1	[1,1'-Biphenyl]-4,4'-diamine, 3,3'-dichloro-
U073	91-94-1	3,3'-Dichlorobenzidine
U074	764-41-0	1,4-Dichloro-2-butene (I,T)
U074	764-41-0	2-Butene, 1,4-dichloro- (I,T)
U075	75-71-8	Dichlorodifluoromethane
U075	75-71-8	Methane, dichlorodifluoro-
U076	75-34-3	Ethane, 1,1-dichloro-
U076	75-34-3	Ethylidene dichloride
U077	107-06-2	Ethane, 1,2-dichloro-
U077	107-06-2	Ethylene dichloride
U078	75-35-4	1,1-Dichloroethylene
U078	75-35-4	Ethene, 1,1-dichloro-
U079	156-60-5	1,2-Dichloroethylene
U079	156-60-5	Ethene, 1,2-dichloro-, (E)-
U080	75-09-2	Methane, dichloro-
U080	75-09-2	Methylene chloride
U081	120-83-2	2,4-Dichlorophenol
U081	120-83-2	Phenol, 2,4-dichloro-
U082	87-65-0	2,6-Dichlorophenol
U082	87-65-0	Phenol, 2,6-dichloro-
U083	78-87-5	Propane, 1,2-dichloro-
U083	78-87-5	Propylene dichloride
U084	542-75-6	1,3-Dichloropropene
U084	542-75-6	Propene, 1,3-dichloro-
U085	1464-53-5	1,2:3,4-Diepoxybutane (I,T)
U085	1464-53-5	2,2'-Bioxirane

Table Appendix 1-2. (*continued*)

EPA Haz. Waste #	CAS* Number	Substance	EPA Haz. Waste #	CAS* Number	Substance
U086	1615-80-1	Hydrazine, 1,2-diethyl-	U098	57-14-7	1,1-Dimethylhydrazine
U086	1615-80-1	N,N′-Diethylhydrazine	U098	57-14-7	Hydrazine, 1,1-dimethyl-
U087	3288-58-2	O,O-Diethyl S-methyl dithiophosphate	U099	540-73-8	1,2-Dimethylhydrazine
U087	3288-58-2	Phosphorodithioic acid, O,O-diethyl S-methyl ester	U099	540-73-8	Hydrazine, 1,2-dimethyl-
U088	84-66-2	1,2-Benzenedicarboxylic acid, diethyl ester	U101	105-67-9	2,4-Dimethylphenol
U088	84-66-2	Diethyl phthalate	U101	105-67-9	Phenol, 2,4-dimethyl-
U089	56-53-1	Diethylstilbesterol	U102	131-11-3	1,2-Benzenedicarboxylic acid, dimethyl ester
U089	56-53-1	Phenol, 4,4′-(1,2-diethyl-1,2-ethenediyl)bis-, (E)-	U102	131-11-3	Dimethyl phthalate
U090	94-58-6	1,3-Benzodioxole, 5-propyl-	U103	77-78-1	Dimethyl sulfate
U090	94-58-6	Dihydrosafrole	U103	77-78-1	Sulfuric acid, dimethyl ester
U091	119-90-4	[1,1′-Biphenyl]-4,4′-diamine, 3,3′-dimethoxy-	U105	121-14-2	2,4-Dinitrotoluene
U091	119-90-4	3,3′-Dimethoxybenzidine	U105	121-14-2	Benzene, 1-methyl-2,4-dinitro-
U092	124-40-3	Dimethylamine (I)	U106	606-20-2	2,6-Dinitrotoluene
U092	124-40-3	Methanamine, N-methyl- (I)	U106	606-20-2	Benzene, 2-methyl-1,3-dinitro-
U093	60-11-7	Benzenamine, N,N-dimethyl-4-(phenylazo)-	U107	117-84-0	1,2-Benzenedicarboxylic acid, dioctyl ester
U093	60-11-7	p-Dimethylaminoazobenzene	U107	117-84-0	Di-n-octyl phthalate
U094	57-97-6	7,12-Dimethylbenz[a]anthracene	U108	123-91-1	1,4-Diethyleneoxide
U094	57-97-6	Benz[a]anthracene, 7,12-dimethyl-	U108	123-91-1	1,4-Dioxane
U095	119-93-7	[1,1′-Biphenyl]-4,4′-diamine, 3,3′-dimethyl-	U109	122-66-7	1,2-Diphenylhydrazine
U095	119-93-7	3,3′,-Dimethylbenzidine	U109	122-66-7	Hydrazine, 1,2-diphenyl-
U096	80-15-9	alpha,alpha-Dimethylbenzylhydroperoxide (R)	U110	142-84-7	1-Propanamine, N-propyl- (I)
U096	80-15-9	Hydroperoxide, 1-methyl-1-phenylethyl- (R)	U110	142-84-7	Dipropylamine (I)
U097	79-44-7	Carbamic chloride, dimethyl-	U111	621-64-7	1-Propanamine, N-nitroso-N-propyl-
U097	79-44-7	Dimethylcarbamoyl chloride	U111	621-64-7	Di-n-propylnitrosamine
			U112	141-78-6	Acetic acid, ethyl ester (I)
			U112	141-78-6	Ethyl acetate (I)
			U113	140-88-5	2-Propenoic acid, ethyl ester (I)

Table Appendix 1-2. (*continued*)

EPA Haz. Waste #	CAS* Number	Substance	EPA Haz. Waste #	CAS* Number	Substance
U113	140-88-5	Ethyl acrylate (I)	U128	87-68-3	1,3-Butadiene, 1,1,2,3,4,4-hexachloro-
U114	111-54-6*	Carbamodithioic acid, 1,2-ethanediylbis-, salts & esters	U128	87-68-3	Hexachlorobutadiene
U114	111-54-6*	Ethylenebisdithiocarbamic acid, salts & esters	U129	58-89-9	Cyclohexane, 1,2,3,4,5,6-hexachloro-, (1alpha,2alpha,3beta,4alpha,5alpha, 6beta)-
U115	75-21-8	Ethylene oxide (I,T)	U129	58-89-9	Lindane
U115	75-21-8	Oxirane (I,T)	U130	77-47-4	1,3-Cyclopentadiene, 1,2,3,4,5,5-hexachloro-
U116	96-45-7	2-Imidazolidinethione	U130	77-47-4	Hexachlorocyclopentadiene
U116	96-45-7	Ethylenethiourea	U131	67-72-1	Ethane, hexachloro-
U117	60-29-7	Ethane, 1,1´-oxybis- (I)	U131	67-72-1	Hexachloroethane
U117	60-29-7	Ethyl ether (I)	U132	70-30-4	Hexachlorophene
U118	97-63-2	2-Propenoic acid, 2-methyl-, ethyl ester	U132	70-30-4	Phenol, 2,2´-methylenebis[3,4,6-trichloro-
U118	97-63-2	Ethyl methacrylate	U133	302-01-2	Hydrazine (R,T)
U119	62-50-0	Ethyl methanesulfonate	U134	7664-39-3	Hydrofluoric acid (C,T)
U119	62-50-0	Methanesulfonic acid, ethyl ester	U134	7664-39-3	Hydrogen fluoride (C,T)
U120	206-44-0	Fluoranthene	U135	7783-06-4	Hydrogen sulfide
U121	75-69-4	Methane, trichlorofluoro-	U135	7783-06-4	Hydrogen sulfide H_2S
U121	75-69-4	Trichloromonofluoromethane	U136	75-60-5	Arsinic acid, dimethyl-
U122	50-00-0	Formaldehyde	U136	75-60-5	Cacodylic acid
U123	64-18-6	Formic acid (C,T)	U137	193-39-5	Indeno[1,2,3-cd]pyrene
U124	110-00-9	Furan (I)	U138	74-88-4	Methane, iodo-
U124	110-00-9	Furfuran (I)	U138	74-88-4	Methyl iodide
U125	98-01-1	2-Furancarboxaldehyde (I)	U140	78-83-1	1-Propanol, 2-methyl- (I,T)
U125	98-01-1	Furfural (I)	U140	78-83-1	Isobutyl alcohol (I,T)
U126	765-34-4	Glycidylaldehyde	U141	120-58-1	1,3-Benzodioxole, 5-(1-propenyl)-
U126	765-34-4	Oxiranecarboxyaldehyde	U141	120-58-1	Isosafrole
U127	118-74-1	Benzene, hexachloro-	U142	143-50-0	1,3,4-Metheno-2H-cyclobuta[cd] pentalen-2-one, 1,1a,3,3a,4,5,5,5a,5b, 6-decachlorooctahydro-
U127	118-74-1	Hexachlorobenzene			

Table Appendix 1-2. (*continued*)

EPA Haz. Waste #	CAS* Number	Substance	EPA Haz. Waste #	CAS* Number	Substance
U142	143-50-0	Kepone	U155	91-80-5	Methapyrilene
U143	303-34-4	2-Butenoic acid, 2-methyl-, 7-[[2,3-dihydroxy-2-(1-methoxyethyl)-3-methyl-1-oxobutoxy]methyl]-2,3,5,7a-tetrahydro-1H-pyrrolizin-1-yl ester, [1S-[1alpha(Z),7(2S*,3R*),7aalpha]]-	U156	79-22-1	Carbonochloridic acid, methyl ester (I,T)
			U156	79-22-1	Methyl chlorocarbonate (I,T)
			U157	56-49-5	3-Methylcholanthrene
U143	303-34-4	Lasiocarpine	U157	56-49-5	Benz[j]aceanthrylene, 1,2-dihydro-3-methyl-
U144	301-04-2	Acetic acid, lead(2+) salt			
U144	301-04-2	Lead acetate	U158	101-14-4	4,4'-Methylenebis(2-chloroaniline)
U145	7446-27-7	Lead phosphate	U158	101-14-4	Benzenamine, 4,4'-methylenebis[2-chloro-
U145	7446-27-7	Phosphoric acid, lead(2+) salt (2:3)	U159	78-93-3	2-Butanone (I,T)
U146	1335-32-6	Lead subacetate	U159	78-93-3	Methyl ethyl ketone (MEK) (I,T)
U146	1335-32-6	Lead, bis(acetato-O)tetrahydroxytri-	U160	1338-23-4	2-Butanone, peroxide (R,T)
U147	108-31-6	2,5-Furandione	U160	1338-23-4	Methyl ethyl ketone peroxide (R,T)
U147	108-31-6	Maleic anhydride	U161	108-10-1	4-Methyl-2-pentanone (I)
U148	123-33-1	3,6-Pyridazinedione, 1,2-dihydro-	U161	108-10-1	Methyl isobutyl ketone (I)
U148	123-33-1	Maleic hydrazide	U161	108-10-1	Pentanol, 4-methyl-
U149	109-77-3	Malononitrile	U162	80-62-6	2-Propenoic acid, 2-methyl-, methyl ester (I,T)
U149	109-77-3	Propanedinitrile	U162	80-62-6	Methyl methacrylate (I,T)
U150	148-82-3	L-Phenylalanine, 4-[bis(2-chloroethyl)amino]-	U163	70-25-7	Guanidine, N-methyl-N'-nitro-N-nitroso-
U150	148-82-3	Melphalan	U163	70-25-7	MNNG
U151	7439-97-6	Mercury	U164	56-04-2	4(1H)-Pyrimidinone, 2,3-dihydro-6-methyl-2-thioxo-
U152	126-98-7	2-Propenenitrile, 2-methyl- (I,T)	U164	56-04-2	Methylthiouracil
U152	126-98-7	Methacrylonitrile (I,T)	U165	91-20-3	Naphthalene
U153	74-93-1	Methanethiol (I,T)	U166	130-15-4	1,4-Naphthalenedione
U153	74-93-1	Thiomethanol (I,T)	U166	130-15-4	1,4-Naphthoquinone
U154	67-56-1	Methanol (I)	U167	134-32-7	1-Naphthalenamine
U154	67-56-1	Methyl alcohol (I)	U167	134-32-7	alpha-Naphthylamine
U155	91-80-5	1,2-Ethanediamine, N,N-dimethyl-N'-2-pyridinyl-N'-(2-thienylmethyl)-			

Table Appendix 1-2. (*continued*)

EPA Haz. Waste #	CAS* Number	Substance	EPA Haz. Waste #	CAS* Number	Substance
U168	91-59-8	2-Naphthalenamine	U183	603-93-5	Pentachlorobenzene
U168	91-59-8	beta-Naphthylamine	U183	608-93-5	Benzene, pentachloro-
U169	98-95-3	Benzene, nitro-	U184	76-01-7	Ethane, pentachloro-
U169	98-95-3	Nitrobenzene (I,T)	U184	76-01-7	Pentachloroethane
U170	100-02-7	p-Nitrophenol	U185	82-68-8	Benzene, pentachloronitro-
U170	100-02-7	Phenol, 4-nitro-	U185	82-68-8	Pentachloronitrobenzene (PCNB)
U171	79-46-9	2-Nitropropane (I,T)	U186	504-60-9	1-Methylbutadiene (I)
U171	79-46-9	Propane, 2-nitro- (I,T)	U186	504-60-9	1,3-Pentadiene (I)
U172	924-16-3	1-Butanamine, N-butyl-N-nitroso-	U187	62-44-2	Acetamide, N-(4-ethoxyphenyl)-
U172	924-16-3	N-Nitrosodi-n-butylamine	U187	62-44-2	Phenacetin
U173	1116-54-7	Ethanol, 2,2´-(nitrosoimino)bis-	U188	108-95-2	Phenol
U173	1116-54-7	N-Nitrosodiethanolamine	U189	1314-80-3	Phosphorus sulfide (R)
U174	55-18-5	Ethanamine, N-ethyl-N-nitroso-	U189	1314-80-3	Sulfur phosphide (R)
U174	55-18-5	N-Nitrosodiethylamine	U190	85-44-9	1,3-Isobenzofurandione
U176	759-73-9	N-Nitroso-N-ethylurea	U190	85-44-9	Phthalic anhydride
U176	759-73-9	Urea, N-ethyl-N-nitroso-	U191	109-06-8	2-Picoline
U177	684-93-5	N-Nitroso-N-methylurea	U191	109-06-8	Pyridine, 2-methyl-
U177	684-93-5	Urea, N-methyl-N-nitroso-	U192	23950-58-5	Benzamide, 3,5-dichloro-N-(1,1-dimethyl-2-propynyl)-
U178	615-53-2	Carbamic acid, methylnitroso-, ethyl ester	U192	23950-58-5	Pronamide
U178	615-53-2	N-Nitroso-N-methylurethane	U193	1120-71-4	1,2,-Oxathiolane, 2,2-dioxide
U179	100-75-4	N-Nitrosopiperidine	U193	1120-71-4	1,3-Propane sultone
U179	100-75-4	Piperidine, 1-nitroso-	U194	107-10-8	1-Propanamine (I,T)
U180	930-55-2	N-Nitrosopyrrolidine	U194	107-10-8	n-Propylamine (I,T)
U180	930-55-2	Pyrrolidine, 1-nitroso-	U196	110-86-1	Pyridine
U181	99-55-8	5-Nitro-o-toluidine	U197	106-51-4	2,5-Cyclohexadiene-1,4-dione
U181	99-55-8	Benzenamine, 2-methyl-5-nitro-	U197	106-51-4	p-Benzoquinone
U182	123-63-7	1,3,5-Trioxane, 2,4,6-trimethyl-	U200	50-55-5	Reserpine
U182	123-63-7	Paraldehyde			

Table Appendix 1-2. (*continued*)

EPA Haz. Waste #	CAS* Number	Substance	EPA Haz. Waste #	CAS* Number	Substance
U200	50-55-5	Yohimban-16-carboxylic acid, 11,17-dimethoxy-18-[(3,4,5-trimethoxy benzoyl)oxy]-, methyl ester, (3beta, 16beta,17alpha,18beta,20alpha)-	U214	563-68-8	Thallium(I) acetate
			U215	6533-73-9	Carbonic acid, dithallium(1+) salt
			U215	6533-73-9	Thallium(I) carbonate
U201	108-46-3	1,3-Benzenediol	U216	7791-12-0	Thallium chloride TlCl
U201	108-46-3	Resorcinol	U216	7791-12-0	Thallium(I) chloride
U203	94-59-7	1,3-Benzodioxole, 5-(2-propenyl)-	U217	10102-45-1	Nitric acid, thallium(1+) salt
U203	94-59-7	Safrole	U217	10102-45-1	Thallium(I) nitrate
U204	7783-00-8	Selenious acid	U218	62-55-5	Ethanethioamide
U204	7783-00-8	Selenium dioxide	U218	62-55-5	Thioacetamide
U205	7488-56-4	Selenium sulfide	U219	62-56-6	Thiourea
U205	7488-56-4	Selenium sulfide SeS$_2$(R,T)	U220	108-88-3	Benzene, methyl-
U206	18883-66-4	D-Glucose, 2-deoxy-2-[[(methylnitroso amino)-carbonyl]amino]-	U220	108-88-3	Toluene
			U221	25376-45-8	Benzenediamine, ar-methyl-
U206	18883-66-4	Glucopyranose, 2-deoxy-2-(3-methyl-3-nitrosoureido)-, D-	U221	25376-45-8	Toluenediamine
			U222	636-21-5	Benzenamine, 2-methyl-, hydrochloride
U206	18883-66-4	Streptozotocin	U222	636-21-5	o-Toluidine hydrochloride
U207	95-94-3	1,2,4,5-Tetrachlorobenzene	U223	26471-62-5	Benzene, 1,3,-diisocyanatomethyl-(R,T)
U207	95-94-3	Benzene, 1,2,4,5-tetrachloro-			
U208	630-20-6	1,1,1,2-Tetrachloroethane	U223	26471-62-5	Toluene diisocyanate (R,T)
U208	630-20-6	Ethane, 1,1,1,2-tetrachloro-	U225	75-25-2	Bromoform
U209	79-34-5	1,1,2,2-Tetrachloroethane	U225	75-25-2	Methane, tribromo-
U209	79-34-5	Ethane, 1,1,2,2-tetrachloro-	U226	71-55-6	1,1,1-Trichloroethane
U210	127-18-4	Ethene, tetrachloro-	U226	71-55-6	Ethane, 1,1,1-trichloro-
U210	127-18-4	Tetrachloroethylene	U226	71-55-6	Methyl chloroform
U211	56-23-5	Carbon tetrachloride	U227	79-00-5	1,1,2-Trichloroethane
U211	56-23-5	Methane, tetrachloro-	U227	79-00-5	Ethane, 1,1,2-trichloro-
U213	109-99-9	Furan, tetrahydro- (I)	U228	79-01-6	Ethene, trichloro-
U213	109-99-9	Tetrahydrofuran (I)	U228	79-01-6	Trichloroethylene
U214	563-68-8	Acetic acid, thallium(1+) salt			

Table Appendix 1-2. (*continued*)

EPA Haz. Waste #	CAS* Number	Substance
U234	99-35-4	1,3,5-Trinitrobenzene (R,T)
U234	99-35-4	Benzene, 1,3,5-trinitro-
U235	126-72-7	1-Propanol, 2,3-dibromo-, phosphate (3:1)
U235	126-72-7	Tris(2,3-dibromopropyl) phosphate
U236	72-57-1	2,7-Naphthalenedisulfonic acid, 3,3′-[(3,3′-dimethyl[1,1′-biphenyl]-4,4′-diyl)-bis(azo)bis[5-amino-4-hydroxy]-, tetrasodium salt
U236	72-57-1	Trypan blue
U237	66-75-1	2,4-(1H,3H)-Pyrimidinedione, 5-[bis(2-chloroethyl)amino]-
U237	66-75-1	Uracil mustard
U238	51-79-6	Carbamic acid, ethyl ester
U238	51-79-6	Ethyl carbamate (urethane)
U239	1330-20-7	Benzene, dimethyl- (I)
U239	1330-20-7	Xylene (I)
U240	94-75-7*	2,4-D, salts & esters
U240	94-75-7*	Acetic acid, (2,4-dichlorophenoxy)-, salts & esters
U243	1888-71-7	1-Propene, 1,1,2,3,3,3-hexachloro-
U243	1888-71-7	Hexachloropropene
U244	137-26-8	Thioperoxydicarbonic diamide [(H$_2$N)C(S)]$_2$S$_2$, tetramethyl-
U244	137-26-8	Thiram
U246	506-68-3	Cyanogen bromide (CN)Br
U247	72-43-5	Benzene, 1,1′-(2,2,2-trichloro ethylidene)bis[4-methoxy-
U247	72-43-5	Methoxychlor
U248	81-81-2*	2H-1-Benzopyran-2-one, 4-hydroxy-3-(3-oxo-1-phenyl-butyl)-, & salts, when present at concentrations of 0.3% or less
U248	81-81-2*	Warfarin, & salts, when present at concentrations of 0.3% or less
U249	1314-84-7	Zinc phosphide Zn$_3$P$_2$, when present at concentrations of 10% or less
U271	17804-35-2	Benomyl
U271	17804-35-2	Carbamic acid, [1-[(butylamino)carbonyl]-1H-benzimidazol-2-yl]-, methyl ester
U278	22781-23-3	1,3-Benzodioxol-4-ol, 2,2-dimethyl-, methyl carbamate
U278	22781-23-3	Bendiocarb
U279	63-25-2	1-Naphthalenol, methylcarbamate
U279	63-25-2	Carbaryl
U280	101-27-9	Barban
U280	101-27-9	Carbamic acid, (3-chlorophenyl)-, 4-chloro-2-butynyl ester
U328	95-53-4	Benzenamine, 2-methyl-
U328	95-53-4	o-Toluidine
U353	106-49-0	Benzenamine, 4-methyl-
U353	106-49-0	p-Toluidine
U359	110-80-5	Ethanol, 2-ethoxy-
U359	110-80-5	Ethylene glycol monoethyl ether
U364	22961-82-6	1,3-Benzodioxol-4-ol, 2,2-dimethyl-,
U364	22961-82-6	Bendiocarb phenol
U367	1563-38-8	7-Benzofuranol, 2,3-dihydro-2,2-dimethyl-
U367	1563-38-8	Carbofuran phenol
U372	10605-21-7	Carbamic acid, 1H-benzimidazol-2-yl, methyl ester
U372	10605-21-7	Carbendazim
U373	122-42-9	Carbamic acid, phenyl-, 1-methylethyl ester

Table Appendix 1-2. (*continued*)

EPA Haz. Waste #	CAS* Number	Substance
U373	122-42-9	Propham
U387	52888-80-9	Carbamothioic acid, dipropyl-, S-(phenylmethyl) ester
U387	52888-80-9	Prosulfocarb
U389	2303-17-5	Carbamothioic acid, bis(1-methylethyl)-, S-(2,3,3-trichloro-2-propenyl) ester
U389	2303-17-5	Triallate
U394	30558-43-1	A2213
U394	30558-43-1	Ethanimidothioic acid, 2-(dimethylamino)-N-hydroxy-2-oxo-, methyl ester
U395	5952-26-1	Diethylene glycol, dicarbamate
U395	5952-26-1	Ethanol, 2,2′-oxybis-, dicarbamate
U404	121-44-8	Ethanamine, N,N-diethyl-
U404	121-44-8	Triethylamine
U409	23564-05-8	Carbamic acid, [1,2-phenylenebis (iminocarbonothioyl)]bis-, dimethyl ester
U409	23564-05-8	Thiophanate-methyl
U410	59669-26-0	Ethanimidothioic acid, N,N′-[thiobis[(methylimino)carbonyloxy]] bis-, dimethyl ester
U410	59669-26-0	Thiodicarb
U411	114-26-1	Phenol, 2-(1-methylethoxy)-, methylcarbamate
U411	114-26-1	Propoxur
See F027	93-76-5	Acetic acid, (2,4,5-trichlorophenoxy)-
See F027	87-86-5	Pentachlorophenol
See F027	58-90-2	Phenol, 2,3,4,6-tetrachloro-
See F027	95-95-4	Phenol, 2,4,5-trichloro-
See F027	88-06-2	Phenol, 2,4,6-trichloro-
See F027	87-86-5	Phenol, pentachloro-

EPA Haz. Waste #	CAS* Number	Substance
See F027	93-72-1	Propanoic acid, 2-(2,4,5-trichlorophenoxy)-
See F027	93-72-1	Silvex (2,4,5-TP)
See F027	58-90-2	2,3,4,6-Tetrachlorophenol
See F027	93-76-5	2,4,5-T
See F027	95-95-4	2,4,5-Trichlorophenol
See F027	88-06-2	2,4,6-Trichlorophenol

Notes

*CAS Number: Given for parent compound only.

NA: Not applicable.

P-listed Waste: The primary hazardous properties of these materials is indicated by the letters T (toxicity) and R (reactivity). Absence of a letter indicates that the compound only is listed for acute toxicity.

U-listed Waste: The primary hazardous properties of these materials is indicated by the letters T (toxicity), R (reactivity), I (ignitability), and C (corrosivity). Absence of a letter indicates that the compound is only listed for toxicity.

Source: http://ecfr.gpoaccess.gov/cgi/t/text/text-idx?c=ecfr&sid=f3fc1aeedcb0cc87bf4367f0a9cc03d0&rgn=div5&view=text&node=40:26.0.1.1.2&idno=40#40:26.0.1.1.2.4.1.4

Appendix 2:
For Further Information

- **RCRA/Superfund Hotline**
  From 1982 to 2004, EPA offered one of the most useful resources available to the hazardous waste community, first known as the RCRA Call Center and then the RCRA/Superfund Hotline, then finally the RCRA, Superfund & EPCRA Call Center. It was dissolved due to lack of funding. However, some of the information gathered by the call center lives in RCRA Online at http://www.epa.gov/epawaste/inforesources/online/ and training modules at http://www.epa.gov/wastes/inforesources/pubs/rmods.htm.

- *Code of Federal Regulations*
  You can still buy the CFR at U.S. government bookstores or access continuously updated regulations, called the e-CFR, at http://ecfr.gpoaccess.gov.

- **EPA Guidance Documents**
  SW-846 and many listing background documents are available online. http://www.epa.gov/epawaste/hazard/testmethods/sw846/index.htm and http://www.epa.gov/lawsregs/policy/sgd/.

- **State Environmental Agency**
  Information on state solid and hazardous waste programs is available at http://www.epa.gov/epawaste/wyl/stateprograms.htm.

- **EPA Headquarters: Office of Resource Conservation and Recovery**
  The office formerly known as OSW (Office of Solid Waste) is now "Office of Resource Conservation and Recovery." Its website is http://www.epa.gov/osw/index.htm.

- **Regulatory Clarifications**
  The regulatory clarifications, once available as the "RCRA Permit Policy Compendium," are now online through RCRA Online at http://www.epa.gov/epawaste/inforesources/online/.

APPENDIX 3:
Significant Regulatory Memos

- **Spent Solvents**
  Doc #9441.1996(13)
  December 24, 1996

  Adding a listed solvent (F001–F005) to a product, such as for thinning paint, is not "use as a solvent" for purposes of characterizing spent solvents listings F001–F005.

  http://yosemite.epa.gov/osw/rcra.nsf/ea6e50dc6214725285256bf00063269d/DC93524D6942FAD28525670F006C2894/$file/14005.pdf

- **RCRA Status of Contaminated Rags**
  Doc #9441.1994(02)
  February 14, 1994

  Rags can be defined as hazardous if they contain or are mixed with listed hazardous waste.

  http://yosemite.epa.gov/osw/rcra.nsf/ea6e50dc6214725285256bf00063269d/7E21AC5A33205BBD8525670F006BEE2F/$file/11813.pdf

- **Aerosol Cans as Scrap Metal**
  Doc #9442.1993(02)
  October 7, 1993

  Aerosol cans that do not contain significant amounts of liquid, exhibit a hazardous characteristic, but will be reclaimed for metal value are exempt from regulation as hazardous waste under §261.6(c).

  http://yosemite.epa.gov/osw/rcra.nsf/ea6e50dc6214725285256bf00063269d/0C95B3D30E33CDB68525670F006BECE7/$file/11782.pdf

■ **Corrosivity Characteristic**
Doc #9443.1993(05)
April 23, 1993

For purposes of the corrosivity characteristic, "aqueous" means any physical form in which pH can be measured.

http://yosemite.epa.gov/osw/rcra.nsf/ea6e50dc6214725285256bf00063269d/
E755CA04BA25BD118525670F006BEB2A/$file/11738.pdf

■ **Non-listed Commercial Chemical Products**
Doc #9444.1993(01)
February 23, 1993

For purposes of the definition of solid waste (§261.2), "commercial chemical product" includes chemical products that are not listed in §261.33 but exhibit one or more hazardous characteristics.

http://yosemite.epa.gov/osw/rcra.nsf/ea6e50dc6214725285256bf00063269d/
D5DB9C1105A9D6858525670F006BEAAB/$file/11726.pdf

■ **Contained-In Policy**
Doc #9441.1989(30)
June 19, 1989

Describes "contained-in" for mixtures of listed hazardous wastes with environmental media.

http://yosemite.epa.gov/osw/rcra.nsf/ea6e50dc6214725285256bf00063269d/
651F2340038DCB1E8525670F006BDEC4/$file/11434.pdf

■ **Recycling/Waste as Raw Material**
Doc #9441.1989(19)
April 26, 1989

This letter is often still used to help with waste vs. raw material decisions.

http://yosemite.epa.gov/osw/rcra.nsf/0c994248c239947e85256d090071175f/
9F219844C3887C378525670F006BDE79/$file/11426.pdf

■ **Aerosol Cans as Reactive**
Doc #9441.1987(77)
September 1, 1987

Regardless of whether an aerosol can is "empty" as defined by §261.7, it may exhibit reactivity due the pressure of residual propellant in the can by being "capable of detonation or explosive reaction" if subjected to heat, pressure, or puncture.

http://yosemite.epa.gov/osw/rcra.nsf/ea6e50dc6214725285256bf00063269d/
1CDB490F11396131852569C900623DC9/$file/13027.pdf

■ **Spent Solvents**
Doc #9441.1986(92)
December 5, 1986

Spent solvents listings (F001–F005) do not apply to wastes resulting from using those listed chemicals as feedstocks.

http://yosemite.epa.gov/osw/rcra.nsf/ea6e50dc6214725285256bf00063269d/
2B96562F26FC7D878525670F006BF997/$file/12809.pdf

■ **Contained-In Policy**
Doc #9441.1986(83)
November 13, 1986

Describes EPA's "contained-in policy" as it applies to groundwater contaminated with hazardous waste leachate.

http://yosemite.epa.gov/osw/rcra.nsf/ea6e50dc6214725285256bf00063269d/
3FEF54CA6D05C4D38525670F006BD5A9/$file/11195.pdf

■ **Using Total Constituent Analysis Instead of TCLP Test**
Doc #9451.1986(03)
April 28, 1986

Rather than conduct a TCLP leachate test, generators may use total constituent waste data to determine if a waste cannot exhibit the toxicity characteristic.

http://yosemite.epa.gov/osw/rcra.nsf/0c994248c239947e85256d090071175f/
7D677DD1D04440B68525670F006C1369/$file/12630.pdf

■ **Tank Cars as RCRA Empty**
Doc #9441.1986(02)
January 7, 1986

Definition of container and the "empty" criteria of §261.7 apply to tank cars.

http://yosemite.epa.gov/osw/rcra.nsf/0c994248c239947e85256d090071175f/
1735466E0E85FCDA8525670F006C1530/$file/12534.pdf

■ **Residue in Empty Containers**
Doc #9441.1984(25)
September 10, 1984

Residue in a container that is empty according to §261.7 is not hazardous waste. If the residue is removed, it is hazardous waste only if it exhibits a characteristic.

http://yosemite.epa.gov/osw/rcra.nsf/ea6e50dc6214725285256bf00063269d/
D8340283232006638525670F006BFDF6/$file/12299.pdf

Index

ORDER FORM

How to Recognize a Hazardous Waste (even if it's wearing dark glasses)

12th Edition | ISBN-13: 978-0-9817753-2-6

Volume Discount Direct from Publisher | 5 or More Copies | U.S. Only

(for individual copies, please go to www.amazon.com)

---

1. Shipping Address

_____

Name

_____

Number, Street, and Apt. Number – or – PO Box Number

_____

City   State   ZIP Code

2. Contact Information

_____

Daytime Telephone   Email Address

3. Quantity of Books Ordered (minimum order of 5 books)

Please mail _____ books to the shipping address provided above.

I have enclosed a ☐ cashier's check or ☐ money order for $_____ to pay for the books.

Pricing (includes shipping and handling via UPS)

Books: 5 through 10 @ $25 each | 11 through 20 @ $22 each | 20 and above @ $18 each

Example 1: 15 books = (10 × $25) + (5 × $22) = $250 + $110 = $360

Example 2: 30 books = (10 × $25) + (10 × $22) + (10 × $18) = $250 + $220 + $180 = $650

Example 3: 100 books = (10 × $25) + (10 × $22) + (80 × $18) = $250 + $220 + $1,440 = $1,910

Make cashier's check or money order payable to OLAP World Press and send it with the order form to:

OLAP World Press
114 Elysian Street
Pittsburgh, PA 15206

Convenient online price calculator (or order online using PayPal or credit card) at:

www.OLAPWorldPress.com/DarkGlasses

www.ingramcontent.com/pod-product-compliance
Lightning Source LLC
Chambersburg PA
CBHW080548220326
41599CB00032B/6410